MULTIMEDIA COMMUNICATIONS

Protocols and Applications

Edited by:
Franklin F. Kuo
Wolfgang Effelsberg
J.J. Garcia-Luna-Aceves

To join a Prentice Hall PTR Internet
mailing list, point to:
http://www.prenhall.com/mail_lists/

Prentice Hall PTR
Upper Saddle River, NJ 07458

ISBN 0-13-856923-1

90000

9 780138 569235

Library of Congress Cataloging-in-Publication Data

Multimedia communications : protocols and applications / edited by
    Franklin F. Kuo, Wolfgang Effelsberg, J.J. Garcia-Luna-Aceves.
       p. cm.
    Includes bibliographical references and index.
    ISBN 0-13-856923-1 (cloth : alk. paper)
    1. Multimedia systems.   2. Telecommunication systems.   I. Kuo.
Franklin F.   II. Effelsberg, Wolfgang.   III. Garcia-Luna-Aceves, J. J.
QA76.575.M777   1977
006.7--dc21   9736112
                                CIP

Editorial/production supervision: *Patti Guerrieri*
Cover design director: *Jerry Votta*
Cover designer: *Scott Weiss*
Manufacturing manager: *Alexis R. Heydt*
Marketing manager: *Miles Williams*
Acquisitions editor: *Mary Franz*
Editorial assistant: *Noreen Regina*

 ©1998 by Prentice Hall PTR
Prentice-Hall, Inc.
A Simon & Schuster Company
Upper Saddle River, NJ 07458

Prentice Hall books are widely used by corporations and government agencies
for training, marketing, and resale.

The publisher offers discounts on this book when ordered in bulk quantities.
For more information, contact: Corporate Sales Department, Phone: 800-382-3419;
Fax: 201-236-7141; E-mail: corpsales@prenhall.com; or write: Prentice Hall PTR,
Corp. Sales Dept., One Lake Street, Upper Saddle River, NJ 07458.

AIX is a trademark of International Business Machines, Inc. Java is a trademark of Sun
Microsystems, Inc. Microsoft Incarta and Windows are trademarks of Microsoft Corporation.
Netscape is a registered trademark of Netscape Communications Corporation. PostScript is a
registered trademark of Adobe Corporation. UNIX is a registered trademark of Santa Cruz
Operations, Inc. All other products or services mentioned in this book are the trademarks or service
marks of their respective companies or organizations.

Printed in the United States of America
10 9 8 7 6 5 4 3 2 1

ISBN   0-13-856923-1

Prentice-Hall International (UK) Limited, *London*
Prentice-Hall of Australia Pty. Limited, *Sydney*
Prentice-Hall Canada Inc., *Toronto*
Prentice-Hall Hispanoamericana, S.A., *Mexico*
Prentice-Hall of India Private Limited, *New Delhi*
Prentice-Hall of Japan, Inc., *Tokyo*
Simon & Schuster Asia Pte. Ltd., *Singapore*
Editora Prentice-Hall do Brasil, Ltda., *Rio de Janeiro*

Table of Contents

Authors

Franklin F. Kuo (Chapters 1, 2)
General Wireless Communications, Inc., Santa Clara, CA

Ralf Keller (Chapter 3)
Ericsson Eurolab Deutschland GmbH, Herzogenrath, Germany

Heinrich J. Stüttgen (Chapter 4)
NEC Europe Ltd., Computer and Communication Research Laboratories, Heidelberg, Germany

Wolfgang Effelsberg (Chapter 5)
Praktische Informatik IV, University of Mannheim, Germany

J.J. Garcia-Luna-Aceves (Chapter 6)
Computer Engineering Department, University of California, Santa Cruz, CA

Brian Neil Levine (Chapter 6)
Computer Engineering Department, University of California, Santa Cruz, CA

Torsten Braun (Chapter 7)
IBM European Networking Center, Heidelberg, Germany

Preface

At the dawn of a new millennium, an information revolution is taking place that involves the convergence of communications with computers. The Internet is a first manifestation of that revolution. Soon to come are technologies that will integrate commerce, education, entertainment, and telecommunications. New consumer products are emerging that will combine the functions of the telephone, the personal computer, and television. Radically innovative telecommunications systems are being developed that will enable the free flow of multiple media—voice, data, image, video information—between these new personal information terminals. These new telecommunications systems involve combinations of the switched public telephone network, broadcast and cable-TV (CATV) nets, as well as wide- and local area data networks. In this telecommunications-driven information revolution, the major technology enabler is multimedia.

Who Should Read This Book

The book is a professional reference for electrical engineers and computer scientists. It is also intended as a classroom resource for advanced undergraduate and graduate courses in telecommunications and in computer science.

What This Book Covers

This book presents the basic technical concepts of multimedia technology and the communications principles underlying multimedia networking. It covers the systems aspects of computer communications, centering on the network protocols needed to make multimedia communications practicable. The coverage extends from the lowest (physical) layer protocols to the highest (application) layer.

We also emphasize communications requirements for multimedia, with particular stress on what these requirements imply for the design of network protocols.

In addition to coverage on protocols, we include a comprehensive chapter on the network technology underpinnings that pertain to multimedia communications, including the very important *Asynchronous Transfer Mode (ATM)* technology. Again, systems issues are emphasized, rather than the hardware and software bases of these technologies.

Finally, we examine important applications of multimedia communication and address new systems approaches needed to render these communications as efficient and inexpensive as possible.

Thus, the kinds of applications presented in this book involve both *multimedia* and *communications*.

What This Book Does Not Cover

Not covered here are applications involving the *representation* of multimedia information, such as multimedia authoring and the design of multimedia databases, because they do not necessarily involve communications aspects.

Among other topics not presented in this volume are baseline computer networking principles, such as TCP/IP, routing, and local area networking. It is assumed that the reader has a basic background in computer networking and knows how to use the Internet and browse the World Wide Web.

How the Book Came To Be

The idea for this book resulted from a series of discussions in Mannheim, Germany between two of the editors—Franklin Kuo and Wolfgang Effelsberg. Kuo spent the academic year 1995-96 as a visiting professor at the University of Mannheim and was supported by an Alexander von Humboldt Foundation Research Award. His host was Professor Effelsberg, chair of Praktische Informatik IV (Computer Science) at Uni Mannheim. Kuo was very impressed by the high degree of technological sophistication in multimedia communications that was exhibited not only at the Mannheim center but at other research institutes in Germany, including the IBM European Networking Center in Heidelberg (about 20 kilometers from Mannheim). Kuo and Effelsberg decided to edit a book on multimedia communications, focusing on the expertise and technical insights of European computer scientists working in the field.

Since multimedia is an international technology, a third editor, from the United States, joined the team—J. J. Garcia-Luna-Aceves of the University of California, Santa Cruz. Garcia-Luna-Aceves's expertise on internetworking and reliable multicasting was necessary to make the coverage more complete and up-to-date.

Acknowledgments

The editors would like to express their gratitude to the authors of the chapters. Their spirit of cooperation and willingness to work hard to meet deadlines is greatly appreciated. Special thanks go to Wieland Holfelder at the University of Mannheim who has done an outstanding job as the principal technical editor of this book. Brian Levine of UCSC, one of the authors, provided special assistance to Kuo and Garcia-Luna-Aceves and deserves our special gratitude. We would also like to thank Mary Franz, Executive Editor at Prentice Hall, Professional Technical Reference, for her advice, help, and most of all, her patience. Finally, we wish to acknowledge the generous support of the Alexander von Humboldt Foundation, of Bonn, Germany, which provided the Research Award to Kuo that made the development of this book possible.

Franklin Kuo, Wolfgang Effelsberg, J. J. Garcia-Luna-Aceves

June 1997

Introduction to Multimedia

Franklin Kuo

To most people, multimedia means infor-
mation representation in terms of text, audio, video, etc., but it is more than
that. It connotes not only the ability to represent these various modes of information but also the capability to manipulate and control this information by computer, as well as transporting the information across a telecommunications
channel. Thus, our working definition of multimedia is:

> Multimedia concerns the representation of mixed
> modes of information—text, data, image, audio, and
> video—as digital signals.
> Multimedia communications concerns the technol-
> ogy required to manipulate, transmit, and control these
> audiovisual signals across a networked communications
> channel.

1.1 The Internet and Multimedia Communications

Anyone who has *surfed* the Internet with a World Wide Web (WWW) browser
already has experienced multimedia communications. Most *html* documents on
the net involve both text and images. Some applications also make use of speech,

1

audio, and video. If you have had some experience on the Internet with WWW, you are also painfully aware of the deficiencies of the Internet for multimedia communications. Most of you realize that it takes much longer to download a picture on the Internet than to download text only.

For example, the *home page* of the University of Mannheim shows an image of the Mannheim castle and accompanying text information about the University (Figure 1–1). For local users in Uni-Mannheim using a browser like *Netscape®*,

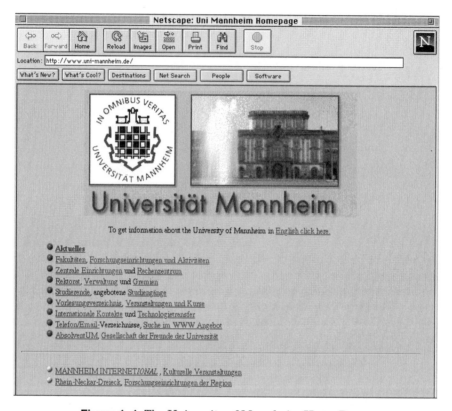

Figure 1–1 The University of Mannheim Home Page

the picture and the text portions show up immediately and simultaneously on the screen. For users in other cities in Germany, the text portion shows up more quickly than the picture. In other parts of the world, the text portion takes much less time to unfold than does the picture portion. In some third-world countries with low-speed links to the Internet, the transmission of graphical images over the Web takes so long as to render it totally impractical.

Why is this so? The simple answer is that pictures or images require much more data space to depict as compared to text. Moving images require even more data space. Before the advent of the WWW, most uses of the Internet involved text- or data-only applications such as e-mail or file transfer (FTP). Real-time communications were not a major concern, and downloading of graphics and images was relatively uncommon. Moreover, the Internet of old was rarely used for speech communications.

The WWW has changed all of that. The Internet today is like an old dirt road which is painfully inadequate for high-speed cars. Nowadays, most countries are planning to build new *information superhighways* for the future Internet, in which multimedia communications are fast and efficient. However, the design of these information superhighways is complex. It does not simply involve building wide-band networks alone. Issues such as access, control, and monitoring must be addressed. These issues and more are the focus of our book.

1.2 Continuous and Discrete Media

In this book, we divide media into two classes: *continuous* and *discrete*. Continuous (also called *temporal*) media change with time. Discrete media are time independent. Examples of continuous media are audio and video. Common examples of discrete media are text, either formatted or unformatted, still images, and graphics.

In this book, we are concerned with both classes of media. It is relatively straightforward to deal with discrete media communications. A text file that is time invariant will not change if sent over the Internet now or ten minutes from now. However, this is not the case with continuous media. A live TV broadcast, sent in real time, contains video and audio media information, both of which are time dependent. The video and audio data streams must be transmitted so that they arrive together, in synchronism. If the communications channel allows the audio stream to proceed more quickly through the channel than through the video stream, the result at the receiving end would be perceptually unpleasing.

Both discrete and continuous media communications place specific requirements on the telecommunications systems used to transport such information, and we examine these requirements in detail.

1.3 Digital Signals

In traditional telecommunications systems, the information transmitted is generally in *analog* form, represented as a continuous (in time) signal. However, in computer communications, the preferred approach is to transmit the audiovisual signals in *digital* form. These digital signals are converted from their original analog representation by two processes: *sampling* and *quantization +encoding*.

1.3.1 Sampling

To sample a signal in time means that the analog signal is examined or measured at regular time intervals T. Thus, a sampled version of a continuous signal $s(t)$ is represented by its values:

$s(t) = \{s(T), s(2T), s(3T),.....s(nT)\}$

where T is the sampling interval, and the sampling frequency is
$f = 1/T$.

According to Nyquist's sampling theorem [1], if a signal contains frequency components up to some frequency f_o, then the minimum sampling frequency that must be employed in order to represent the signal accurately, is $2f_o$. For example, if a speech signal has a maximum frequency of 3 kHz, then the minimum sampling frequency must be at least 6 kHz. That is, the speech signal must be sampled at least 6,000 times per second in order to faithfully represent the original analog speech signal.

1.3.2 Quantization and Encoding

After an analog signal is sampled, the sampled value is then quantized and encoded as a string of bits. Quantization means the representation of the sampled values in terms of a discrete set of amplitude values. The number of bits used to represent a sampled value determines the accuracy of the quantization/coding process. For example, if a sampled value $s(iT)$ is quantized by a 3-bit code, $s(iT)$ is able to assume only 8 different discrete values, from {000, to 111}. Each of these discrete values can then be further encoded to facilitate economies in transmission or storage of the digitized information. In Figure 1–2, we see that an analog signal $s(t)$ is sampled at discrete points in time nT and then quantized and encoded as represented by the dark vertical lines at every sampling interval.

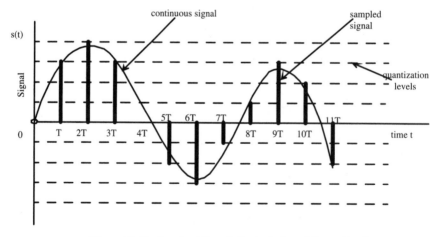

Figure 1–2 Analog Signal, Sampled and Quantized

In telephone speech applications, the current standard for digitizing the human voice is to use 16 bits per sample, which then leads to 2^{16} or 65,536 distinct amplitude levels that could be represented. In other speech compression applications, perhaps only 8 quantization bits are required, which then means that we can only distinguish 256 different amplitudes.

1.3.3 Bit Rate

Bit rate is defined as the product of the sampling rate and the number of bits used in the quantization process. For example, in telephone speech, the bandwidth of the speech signal is around 3 kHz, which means that the sampling rate (frequency), according to Nyquist's theorem, should be at least 6 kHz. Let us assume a more conservative sampling rate of 8 kHz. If we use an 8-bit quantizer, then

the bit rate required for telephone speech = 8000 x 8 = 64 kbps

As another example, consider the case of a compact audio disk (CD). In high fidelity audio, the bandwidth is generally accepted as 20 kHz, so that the minimum sampling frequency is 40 kHz. Using the industry standards of 44.1 kHz sampling rate and 16 bit quantization, and for stereo (2 channels), we arrive at:

the bit rate for compact audio disk = 44100 x 2 x16 = 1,410 kbps

In data communications, bit rate is an important parameter. The channel capacity of public data networks is often given in kilobits or megabits per second. For example in the *Integrated Services Digital Network (ISDN)* the standard bit-rate employed for speech is 64 kbpss.

In Table 1–1 we show the relative bit rates for telephone speech, teleconferenced speech (which requires higher speech quality), compact audio disk (CD), and digital audio tape (DAT).

Format	Sampling rate (kHz)	Bandwidth (kHz)	Frequency Range (Hz)	Bit rate (kbps)
Telephony	8.0	3.0	200 - 3,200	64
Teleconferencing	16.0	7.0	50 - 7,000	256
Compact disk	44.1	20.0	20 -20,000	1,410
Dig Audio Tape	48.0	20.0	20 - 20,000	1,536

Table 1–1 Digital Audio Formats [6]

1.4 Still Images

Earlier, we alluded to the fact that still images take much longer to transmit than text. Why should this be so? The basic reason is that images are composed of *pixels*, and there are a very large number of pixels in each image. What is a pixel? While you might not find this definition in most standard dictionaries, a good technical dictionary might define *pixel* as:

Pixel: (noun) the smallest single unit or point of an image whose color or brightness can be controlled.

If we examine, under a microscope, a photograph on a newspaper page, we see that the photograph is made up of a very fine grid of dots, each dot having a specific tone of gray, ranging from black to white. These dots are pixels, and we see that the pixel can take on a particular value—in the case of a newspaper photo, a tone of gray.

In color pictures or displays, the pixel can also assume values associated with color and brightness.

A better definition of pixel might be:

Pixel: *Picture element.* The smallest resolvable unit area of an image, either on a screen or stored in memory. Each pixel in a monochrome image has its own brightness, from 0 for black to the maximum value (e.g., 255 for an 8-bit pixel) for white. In a color image, each pixel has its own brightness and color.

Computer images are bit maps made up of pixels. In a standard resolution computer display, there are 768 lines, each line containing 1,024 pixels. For a color display, suppose each pixel's color and brightness values is specified by 24 bits (bits per pixel, or bpp), then the total number of bits for an image on the computer screen is

$$\textbf{\textit{Number of bits}} = 1024 \times 768 \times 24 = 18.874 \text{ megabits}$$

Suppose we need to send this image over a 14.4 kilobit/sec modem. It would then require

$$\textbf{\textit{Transmission time}} = 18874000/14400 = 1310 \text{ secs or } 21.84 \text{ min}$$

This is clearly too long for computer users to wait if they are trying to work in real time. What can be done about it? There are four basic approaches:

1. **Send the image over a faster channel**, such as a T1 line with a speed of 1.544 Mbps.
2. **Reduce the number of bits per pixel**, thus allowing fewer discrete levels for brightness and fewer color tones.
3. **Reduce the resolution of the display** (fewer pixels per line and fewer lines per picture).
4. **Remove the redundancy in the display,** which means removing excess pixels that represent essentially the same object.

Image compression techniques combine approaches 2, 3, and 4.

1.5 Text and Graphics

Compared to images, plain text or formatted text requires much less transmission capacity. Plain text characters are represented by 8 bits, or 1 byte. Formatted text characters are represented by 2 bytes. For a single page of text, there are 64 lines, and 80 characters per line. Thus, a single full page of text contains:

$$\textbf{\textit{Number of bits}} = 80 \times 64 \times 2 \times 8 = 82 \text{ Kbits}$$

which requires (using a 14.4 kbps modem) only 5.7 secs to transmit. In contrast to images, graphics are human- or computer-drawn pictures made up of lines in space. We can consider a graphic as a composition of objects that represents information. These objects can be described mathematically or by computer commands. Thus, a straight line can be described in terms of two end points, a circle by the location of its center and its radius, etc. Graphics are revisable, or editable. Images are not. Graphics require much less storage space in memory than does a bit-mapped image. Finally, graphics need much less time to transmit over a network than bit-mapped images need.

In multimedia communications, we need to deal with both images and graphics.

1.6 Moving Graphics and Images

Motion pictures are composed of temporal sequences of graphical pictures, each of which is called a *frame*. The graphical objects in each frame vary slightly from the previous one, so that when the sequence of frames are projected in time, the result shows an object that is perceived to be moving. The speed of the projection in terms of the number of frames per second is called the *frame rate*. If the frame rate is too slow, the result is jerkiness in the motion of the object. Movies usually have a frame rate between 25 and 30 frames per second (fps). Experience shows that a frame rate of 16 or more is needed to depict smooth motion.

Motion picture and video cameras take pictures of real-world moving objects. Each camera shot is an individual frame. A temporal sequence of shots taken over time shows the motion of the objects.

In video displays, the frame rate is that rate at which the temporal sequence is played back. This frame rate is usually 25–30 fps.

How much data is there in a second's worth of video? Consider the case of a common video format, the Common Intermediate Format (CIF), which we discuss in greater depth later in the book. Under the CIF format operating at 30 fps, there are 360 pixels per line, and 288 lines per picture. Each pixel takes on a 24-bit value for brightness and color. Thus, the number of bits in a single second of CIF video is:

   No. of bits per second $= 360 \times 288 \times 24 \times 30 = 74.65$ Mbps

In today's world, the most commonly used multimedia device is the personal computer. And the most commonly used communications device used by PCs is the modem, attached to ordinary telephone lines. Since today's modems have speeds of 28.8 kbps or less, a single second of CIF video requires 2,592 seconds or 43 minutes to transmit. This means that the PC cannot receive the video in real time. It also implies that the PC must have gigabits of video storage memory in order to receive the video data, store it, and play it back some time later.

1.7 Encoding and Decoding

Human beings can sense only audiovisual signals that are analog. For digital communications systems, these analog signals must be converted into digital form by *encoders*, which perform *analog-to-digital (A to D)* conversion as well as quantization and compression. Many modern-day telephone networks, such as the ISDN, are completely digital in that they employ both digital transmission and digital switching. The end users of such telephone systems can be humans or machines (computers). If the sender and receiver are both humans, then the originating and terminating signals must be analog, and A to D and D to A converters must be employed, as shown in Figure 1–3. If, however, the end users are both computers, then such converters are not needed. Aside from D to A and A to

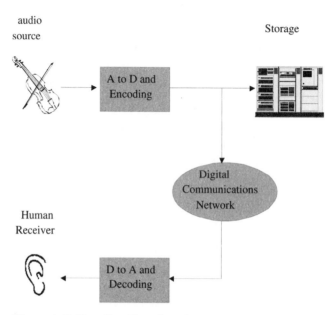

Figure 1–3 Encoding/Decoding for Digital Transmission

D conversion, the encoder/decoder must also perform the function of *data compression,* which connotes the techniques employed to represent the digital signal in a reduced compact form. There are two kinds of compression: *lossless* and *lossy.* For the former, the encoding/decoding process yields a signal at the receiver that is identical to the sending signal. For the latter, there is a difference, which is denoted as *distortion.* The goal of such algorithms is to produce a highly compressed version of the original signal in which the distortion is kept within well-tolerated bounds. In most cases, the compressed signal is a lower bit rate version of the sender's signal.

The use of compression varies according to different media. Speech compression algorithms are markedly different from video compression techniques. The evaluation of which compression methods are better than others depends upon human audiovisual perception, which in turn depends upon perception of signal *quality.* For example, in high-fidelity audio, the frequency range is generally accepted as 20–20,000 Hz. Knowing that most humans cannot hear beyond 16,000 Hz or below 40 Hz, one can devise a compression algorithm which produces acceptable quality audio between these modified frequency bounds. For video, a compression algorithm could be devised in which the number of bits per pixel could be reduced, say, from 24 to 16. Or the frame rate could be reduced from 30 fps to 16 fps in visual scenes where there is no rapid movement.

Many modern compression techniques are derived from the intimate knowledge of the mechanisms of human perception. These techniques are referred to as *perceptual coding*. An example is the concept of *just noticeable distortion (JND)*[7] in audio compression. This is the distortion level at which there is a very small difference between the original and the encoded signal, as perceived by human listeners. For high-fidelity audio, the JND variation versus frequency could be measured or modeled for a wide range of human listeners. With the knowledge of this JND curve, signal components below the curve can be discarded in the compression without the human listener noticing the effects.

1.8 Bandwidth vs. Compression

As we have seen above, signal compression is important for multimedia communications. In today's telecommunications world, considerable progress has been made in both compression technology and high-speed networking. With the greater availability of broader bandwidths at lower costs of local- and wide-area networks, the need for more and more signal compression decreases. However, with more and more users sending and receiving multimedia data, compression is still needed despite the increase in available bandwidths.

Constant progress is being made in compression technology. For example, in telephone speech, the standard bit-rate for network-quality speech was 64 kbps in 1972. In 1984, a new standard at 32 kbps was adopted; and in 1992, yet a newer 16 kbps standard emerged. Work is in progress to lower this standard to 8 kbps. Such progress can be attributed to better and better speech compression algorithms, as well as more and more powerful integrated circuits for speech compression.

Many international compression standards in use today, apply to network telephony, audio, image, and video transmission. Some of the standards are listed in Table 1–2.

Another example of compression is in CD-quality audio. As we have seen, 20 kHz sound is sampled at 44.1 kHz, with 16 bits per pixel to yield 1.412 Mbps stereo. By use of a *Perceptual Audio Coder (PAC)*, AT&T Bell Laboratories scientists have demonstrated the capability of broadcasting CD-quality music at 64 kbps [9]. This means that CD-quality sound can now be sent over a basic-rate ISDN channel. In Figure 1–4, we see a chart depicting the current limits of signal compression.

In Europe and in Japan, use of basic-rate ISDN for business and in-home applications is widespread. In the United States, that is becoming so. A lower-speed alternative to basic-rate ISDN is the V.34 modem standard, which offers data bandwidths of up to 33.6 kbps. With bandwidths such as 64- and 28.8 kbps, multimedia communications with data compression is certainly feasible.

In many commercial environments where local area networks (LANs) offer bandwidths of 10 Mbps or more and where WAN-connectivity of T1 (1.536 Mbps)

Standard	Bit rate	Application
G.721	32 kbps	Telephony
G.728	16 kbps	Telephony
G.722	48-64 kbps	Teleconferencing
MPEG-1 (audio)	128-384 kbps	2-channel audio
MPEG-2 (audio)	320 kbps	5-channel audio
JBIG	0.05-0.10 bpp	Binary images
JPEG	0.25-8.0 bpp	Still images
MPEG-1,2 (video)	1-8 Mbps	Video
Px64	64-1,536 kbps	Videoconferences
HDTV	17 Mbps	Advanced TV

Table 1–2 International Standards for Telephony, Audio, and Video [8]

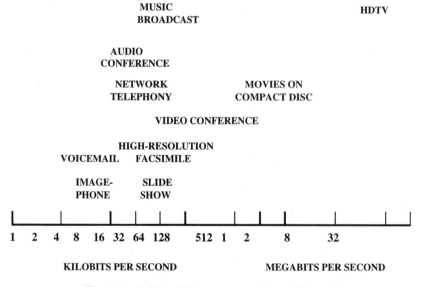

Figure 1–4 Signal Compression Capabilities [7]

rates or greater are available, the use of multimedia is becoming more wide-spread. In the near future, asynchronous transfer mode technology (ATM) will offer higher and higher data rates, both for LANs and WANs. With rapid deployment of ATM-based networks, broadband ISDN, or *B-ISDN*, will be increasingly available for multimedia communications.

In the public-switched telephone network (PSTN), the capability for broadband communications is based upon the use of fiber optics transmission and switching technologies. Currently, the fastest data rate over commercial fiber optics networks is about 2,500 Mbps or 2.5 gigabits per sec (Gbps). In early 1996, Fujitsu Laboratories of Japan, as well as AT&T Bell Laboratories, demonstrated the ability to send data over optical fibers at the rate of 1,100 Gbps or 1.1 Terabits per second. At this speed, 4 million newspaper pages, 250 years' worth of newspapers, can be transmitted in one second. The 1.1 Tbps capacity is equal to 15 million 64 kbps ISDN circuits. It is expected that by the turn of the century, networks based upon this new technology will become widely deployed. By then, multimedia communications will be widespread.

1.9 Project TeleTeaching

As an example of the current applications of multimedia communications, consider the project that one of the authors (Effelsberg) is undertaking in the area of *teleteaching*. At issue is the use of multimedia technology and high-speed networks in broadening the spectrum of courses available to university students and in increasing their comprehension of the proffered course materials. The Universities of Heidelberg and Mannheim in Germany are currently engaged in a joint project testing new technology for distance learning in a digital network. In this project, high-performance multimedia workstations and PCs are connected via ATM, and lectures, exercises, and stored teaching materials are available over the network. University educators and psychologists provide auxiliary scientific evaluation.

Current media-based instructional approaches, such as exchanges of video cassettes or instructional television programs, deprive the *remote student* of any influence upon the content or quality of instruction via questions or contributions to discussions. The instructors, for their part, are deprived of feedback on their presentations and of the students' questions. Project TeleTeaching provides a solution for each of these educational issues. Courses can be exported, enabling interaction between instructors and students. Europe to date has little experience with this form of university teaching.

1.9.1 Background and Motivation

The increasing differentiation and specialization in all fields of knowledge pose problems fundamental not only to university teaching, but to teaching at other levels of education as well. Most university students will come into contact with *authentic* research only within very narrow confines of knowledge. At most universities, it is difficult, if not impossible, to offer courses in every scientific discipline, much less carry out research in every discipline. So it is with the Universities of Heidelberg and Mannheim. Heidelberg has a very distinguished history in the field of physics. However, it does not offer computer science. Mannheim, on the other hand, has a very good computer science department.

Given their geographic proximity (only 20 km apart) and in view of their complementary natures, the Universities of Heidelberg and Mannheim are predestined for cooperative teaching. Mannheim's traditional strengths are in the economics disciplines, whereas liberal arts and natural sciences prevail in Heidelberg. Prompted by this background, both universities signed a far-reaching cooperative agreement in June 1995, in which the two institutions will mutually recognize course credits, and students may use the resources, such as libraries, of each university.

The most important goal of the TeleTeaching project between Heidelberg and Mannheim is an improved availability of courses at both universities. On the one hand, a broader spectrum of courses available at either university is targeted; on the other hand, the use of multimedia teaching materials (animated scenes, visualization of technical calculations, 3-D models, etc.) will enrich and intensify the transfer of knowledge. TeleTeaching's goal is to amplify rather than replace current methods of teaching and learning. The travel saved between Heidelberg and Mannheim, and the resulting reduction in environmental stress, is a welcome benefit.

In particular, the basic courses in computer science at Mannheim will be exported to Heidelberg in exchange for courses in physics This export can be understood as a virtual change of location on the part of the instructors or students, as well as offering an excellent basis for testing teleteaching, the export of courses via a high-speed digital network.

1.9.2 Teleteaching Scenarios

Several plausible scenarios for teleteaching will be tested within this project. In each instance, the latest digital multimedia technology will be used; courses will be exported over an ATM high-speed network. The high bandwidth of ATM enables simultaneous transfer of high-quality video, audio, and data streams in either direction [2][5]. The video channel especially requires high bandwidth and guaranteed *quality of service* characteristics that are not available on the current university networks using LANs with the TCP/IP protocol stack. Thus, the project contributes at the same time to the further development of innovative applications for high-speed networks.

Scenario 1, Auditorium-to-Auditorium

In an initial pilot phase, an auditorium at each university has been equipped with modern audio and video technology and, a multimedia workstation, and has been connected to the other auditorium by means of a digital high-speed network. In these auditoriums, special lectures that are of interest to the partner university will be held and exported live to that partner. Students there will be

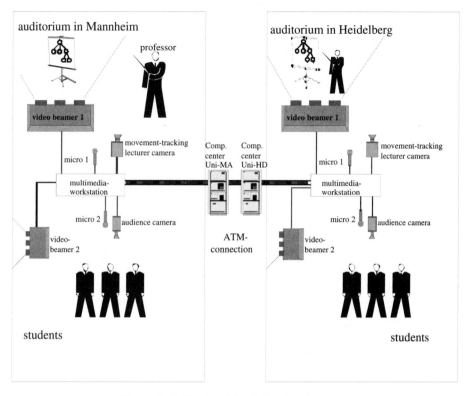

Figure 1–5 TeleteachingInfrastructure

able to ask questions and conduct discussions with the instructor as if they were actually *on location*. This scenario is depicted in Figure 1–5.

The following equipment is installed for the auditorium scenario; the equipment of the auditoriums at either university is equivalent, enabling export in either direction.

- One camera focused on the instructor
- A second camera focused on the student audience.
- A multimedia workstation (rather than an overhead projector), whose screen content is projected on a large screen.

  The transparencies used for the course are produced for the computer, whereby graphics, color pixel scenes, etc., may be used. In the near future, course materials will include moving scenes, particularly animated ones, visualized algorithms and processes, and interactively steerable 3-D models.

- A second projector, which projects, on the side or rear wall of the local auditorium, a view of the audience in the remote auditorium.
- All four cameras with their video and audio streams and the multimedia workstation at the instructor's site are managed by the current instructor. On the part of the instructor, five data streams are activated and exported:

- Video and audio of the instructor's camera
- Video and audio of the local audience's camera
- Blackboard view

These streams are transferred to the remote auditorium site. At the same time, two channels at the remote site are activated: the video and the audio channels of the student audience. The instructor, while lecturing, can see, projected on the side wall of the auditorium at the remote site, himself and the local students. The students in both auditoriums see the projected blackboard view of the multimedia teaching materials.

Transfer of compressed data streams between the two sites is governed by ATM circuits (virtual circuits) activated at a guaranteed data rate in order to preclude any decline in audio and video quality during concurrent transfer of data from other sources of the network.

Scenario 2, Teleteaching Without Being Physically Present

Additional scenarios are planned in the further course of Project TeleTeaching. A futuristic, technically and psychologically interesting example is *computer-to-computer;* it represents the opposite end of the spectrum compared with the auditorium-to-auditorium scenario. Students and the instructor are connected only via the network. The instructor uses the multimedia workstation at his office. He can be connected, for example, to the student's home if high bandwidth access is available. The student sees a video view of the instructor on the computer screen, hears the instructor's voice, and sees in an additional window the *blackboard view,* in which during the lecture, text material, and annotated graphics are presented. The traditional *blackboard view* can also be enriched by means of animated scenes, visualized algorithms, or physical processes described by mathematical systems of equations.

This scenario also permits participation in courses at distant universities. Given an excellent teacher in a specific field elsewhere, the student can *enroll* for the course there. Within just a scant number of years, it will be technologically feasible to participate in courses at MIT or UC Berkeley from a workstation or PC in Mannheim or Heidelberg.

The lack of the physical presence of the instructor poses burning educational and psychological questions. There is no precedent for gauging the learning success in a teleteaching environment compared with that in a traditional teaching environment. How dependent is comprehension of course material upon the physical presence of the instructor? Another unknown variable is how well or poorly the multimedia interaction between student and professor over the network will actually function.

Scenario 3, Network-Linked Tutorial Sessions

Under this scenario, student and instructor terminals are equipped with multimedia computers and high-speed network links. The instructor can request display of the screen contents of each student and interact with each. This scenario resembles that of a language laboratory and is conceived for practice exercises.

Scenario 4, Offline Requests for Course Lectures

Once multimedia technology has been installed, courses can be quite simply recorded (analog) and then digitized and compressed for storage on a large disc server. A video-on-demand server will be created with an index of video recordings suitable for digital remote retrieval. When preparing for examinations, the student can dial up the video-on-demand server and retrieve those sections of the material not yet understood. Ideally, every lecture would be edited after being held. Errors during the lecture or lengthy passages would be edited out of the video. The video would be enhanced by auxiliary reading material in PostScript files, and additional interactive learning materials would be available.

The rewriting, storage, and network-wide transfer of multimedia learning materials is still virgin terrain. Diverse research issues are still unresolved, for example, the automatic indexing of digital videos for optimum distribution and optimum scheduling of videos on distributed servers or for real-time-capable transfer protocols in packet distribution networks [1][3][10].

1.9.3   Multimedia Teaching Materials

Concomitant development of multimedia teaching materials (video clips, animated scenes, interactive simulation programs, etc.) is vital. This is planned for the Project TeleTeaching courses between Mannheim and Heidelberg, constituting the best use of the potential afforded by the new technology. Contributions by the Springer-Verlag publishing house and the Interdisciplinary Institute of Scientific Computing at the University of Heidelberg are important in this regard.

Of particular interest are the simulations operable by the students themselves. For example, a cranial volume model that can be rotated in three-dimensional space and broken down into its component parts can be designed for an anatomy course. With this model, a student's spatial perception of anatomy is much finer than that gained from any anatomy textbook. Or a simulated model of harmonic vibrations for PCs can be designed for a physics course. Both these learning materials are already currently available, and they impart an excellent impression of the potential of the new media in teaching.

The creation of such learning materials involves tremendous effort on the part of instructors, who are unfamiliar with the handling of the new media. While the technical tools of a scientist include preparing written papers or textbooks, and expertise in their use is honed during university studies and doctoral work, the planning and production of a teaching video, an interactive hypermedia document, or interactive learning software poses a new challenge for instructors. Teamwork is one alternative for reducing the workload. Several colleagues holding similar courses at different universities can collaborate, each developing multimedia teaching and learning materials for a relatively small section of the field. Another alternative is *authoring on the fly*. Here, a course is held on a workstation as if before an audience and recorded in audio and video together with the *blackboard view* and annotations. The raw material gained with relatively little effort can be supplemented by written materials and stored on a server for later retrieval.

1.9.4   Industrial Partners

Competent industrial partners in this project include the IBM European Networking Center in Heidelberg, the Deutsche Telekom (German Telecommunications) and the Springer-Verlag publishing house. IBM Heidelberg's multimedia research department enjoys international renown. Telekom is very interested in testing innovative implementations in high-speed networks under realistic conditions. The Springer-Verlag publishing house is the leading developer of multimedia titles and teaching materials in the fields of computer science and physics.

1.9.5   Comparable Projects

The Heidelberg-Mannheim Project is by no means the only one of its kind. Indeed, technology is now developed and ready for teleteaching, as seen in the number of similar activities developing everywhere. The United States and Canada are leaders in the development and testing of teleteaching. The motivation here is often to provide large, sparsely populated states and provinces with high-quality teaching at an acceptable cost. Analog television technology (with feedback channels), a field for experiments, and satellites are used to transmit courses. Many new teleteaching projects using Internet protocols and WWW are underway But there is little literature available on multimedia teaching and learning materials or on the storage of teaching units on video-on-demand servers, on educational development aid, or on a systematic evaluation of learning success. The future lies open, ready and waiting.

References

[1]   C. Bernhardt and E. Biersack, "A Scalable Video Server: Architecture, Design, and Implementation," *Research Memorandum, Institut Eurocom,* Sophia Antipolis, 1994.

[2]   M. de Prycker, *Asynchronous Transfer Mode: Solution for Broadband ISDN,* Ellis Horwood, 1993.

[3]   W. Effelsberg, B. Lamparter, and R. Keller, "Application Layer Issues for Digital Movies in High Speed Networks," In *Architecture and Protocols for High Speed Networks,* edited by O. Spaniol, A. Danthine, and W. Effelsberg. Dordrecht, Kluwer Academic, 1994: 273–292.

[4]   A. Gersho and R. M. Gray, *Vector Quantization and Signal Compression,* Dordrecht: Kluwer Academic, 1992:53.

[5]   R. Haendel and M. N. Huber, *Integrated Broadband Networks: An Introduction to ATM-Based Networks,* Wokingham: Addison Wesley, 1991.

[6]   N. Jayant, B. D. Ackland, V. B. Lawrence, and L. R. Rabiner, "Multimedia: Technology Dimensions and Challenges," *AT&T Technical Journal* 74(5):14–3, September/October 1995.

[7]   N. Jayant, "Signal Compression: Technology Targets and Research Directions," *IEEE Journal on Selected Areas in Communications* 10(5), June 1992.

[8]   N. Jayant, B. D. Ackland, V.B. Lawrence, and L. R. Rabiner., *op cit.*

[9]   N. Jayant, J. Johnston, and R. Safranek "Signal Compression Based on Models of Human Perception," *Proceedings of the IEEE* 81(10), October 1993.

[10] P. V. Rangan and H. M. Vin, "Efficient Storage Techniques for Digital Continuous Media," *IEEE Trans. on Knowledge and Data Engineering,* Aug. 1993.

Multimedia Networks: Requirements and Performance Issues

Franklin Kuo

*I*n this chapter, we define and identify important performance parameters in multimedia networking. We introduce these performance parameters and discuss which roles these parameters play in distributed multimedia communications. More detailed explanations are supplied in subsequent chapters. A simple definition of the word *distributed* is *not in one place*. Distributed multimedia thus connotes the distribution of multimedia information between different geographical locations. Whether these locations are within a single building or are separated by thousands of kilometers does not affect the meaning of *distributed*. In this book, distributed multimedia means transporting multimedia information across a communications network.

2.1 Distributed Multimedia Applications

In recent years, digitization and networking have played ever increasing roles in the evolution towards a distributed information society. Teger [5] illustrates this progression in Figure 2–1.

As noted in Figure 2–1, the applications of distributed multimedia are many and varied. Each of these applications places specific performance requirements on the network. In this chapter, we examine what these requirements are and what they imply about the performance of the networks as distributors of multimedia information.

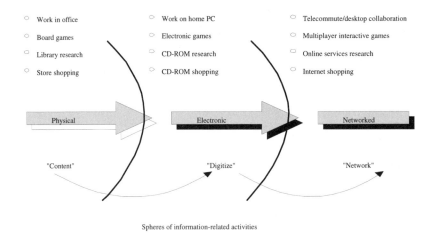

Figure 2–1 Evolution of Networked Services

2.2 Peer-to-Peer and Multipeer Communications

Multimedia communications involve two basic modes: *unicast* and *multicast*. In unicast mode, there are two communicating partners, or *peers*, and the resulting mode is called *peer-to-peer* communications. Multicast mode involves *1 to n communic*ations, or *peer-to-multipeer,* as well as *1 to all* communications or *broadcast* mode.

Unicast applications include individual client-to-server applications with examples such as home-shopping, online banking, video-on-demand, or multimedia e-mail.

In a distance-learning or teleseminar application, where there are one lecturer and multiple remote attendees, the basic mode is that of peer-to-multipeer. Contrast this to a teleconference application, where geographically distributed attendees (peers) interact with each other via multimedia links. In a teleconference, each attendee can assume roles of speaker or listener, so the mode is multipeer-to-multipeer.

A special form of multipeer communications is that of *Multiparty Interactive Multimedia* (MIM)[2], of which an important example is *Computer Supported Collaborative Work*. Under CSCW, geographically distributed co-workers share a multimedia workspace consisting of a common set of files, graphical displays, and a distributed whiteboard. In addition, they share applications such as spreadsheets, editors, and drawing programs. In a CSCW teleconference, collaborative workers are able to solve large design or engineering problems in real time, while working at different locations.

MIM interactions can be classified as either *dynamic* or *static* [2]. Dynamic interactions are those in which all participants are allowed to exchange information at any time. In a static interaction, only a prescribed subset of participants are allowed to present information. A multimedia teleconference, in which all participants are peers, is an example of a dynamic interaction. CSCW is another example. A third example of a dynamic interaction is that of a *Virtual Cafe,* in which users at different sites are allowed to communicate informally through a shared electronic environment [1], much like an Internet *chat* session. The difference between CSCW and a Virtual Cafe is that access to a CSCW session is controlled, whereas access to a Virtual Cafe is not.

In multicast mode, information is passed from a central source to many receivers. This is an example of a static interaction. Another is the so called *monitoring* scenario, in which information from many sources is sent to a single receiver. A third example is that of teleteaching, in which the lecturer is an information source and all other participants are information sinks. Examples of these different interaction modes are shown in Table 2–1.

Type	Description	Interaction	Data Flow	Accessibility
CSCW	All can send/receive data	Dynamic	N to N	Controlled
Virtual Cafe	All can send/receive data	Dynamic	N to N	Uncontrolled
Broadcasting	One sender, many receivers	Static	1 to N	Controlled
Monitoring	One receiver, many senders	Static	N to 1	Controlled

Table 2–1 Characteristics of Selected MIM Applications

2.3 Network Performance Parameters for Multimedia

As we have seen in the previous chapter, there are key network performance parameters for multimedia communications. We have seen that the bit rate of a network is a crucial network characteristic. In this section, we will examine other parameters of importance. They are:

- Throughput
- Error Rate
- Delay

Each of these parameters plays a vital role in enabling the transmission of audiovisual signals over a digital network.

2.3.1 Throughput

The throughput of a network is its effective bit rate, or effective bandwidth. Thus, we define throughput to be the physical-link bit rate minus the various overheads that pertain to the transmission technologies employed. For example, in high-speed networking applications employing ATM technology over a SONET (Synchronous Optical NETwork) fiber optics transmission system, the network carrier's provisioned bit rate is 155.52 Mbps. The principal overheads that pertain are:

- Approximately 3% for SONET
- Approximately 9.5% for ATM

Thus, the maximum throughput of this particular network is about 136 Mbps. In addition to the overhead factors just mentioned, other factors also affect throughput. These include network congestion, bottlenecks, node or line faults, etc.

In many instances, the overhead is deemed to be implicit, and throughput is simply equated to the bit rate of the system. Thus, many people regard the throughput of a basic-rate ISDN system to be 64 kbps, even though the throughput might be somewhat less because of overhead constraints.

2.3.2 Error Rate

Another important parameter for multimedia networks is the error rate. This parameter can be defined in a number of ways. One is *bit error rate (BER)*, which is defined as the ratio of the average number of corrupted or error bits to the total number of bits that are transmitted. Another is *packet error rate, (PER)* defined as above, with *packets* substituted for *bits*. A third parameter is *frame error rate*, which applies to ATM networks, defined as the number of error frames over the total number of frames transmitted. In most of today's networks, bit error rates are very low. For example, in fiber optics transmission, the BERs range from 10^{-9} to 10^{-12}. In satellite transmission systems, the BER is on the order of 10^{-7}. This means that on the average, there is one error bit in a file of ten million bits on a satellite digital circuit. Considering that a single video frame might consist of many millions of bits, such BERs imply that there could be approximately one bit error per frame in digital video transmission.

One might ask if such errors are important. In some applications such as video transmission, a single error in a frame might not be seen by the human eye. In others such as interbank transfer of electronic funds, a single error bit might be catastrophic. In later chapters we will discuss when error rate is of importance and when it is not.

2.3.3 Delay

Delay is one of the most important network performance parameters that we will encounter in this book. It can take many forms. We will discuss this issue from the standpoint of *end-to-end delay*, which means the time it takes to transmit a block of data from the sending to the receiving end system. Components of end-to-end delay are:

- *Transit delay,* which is a physical parameter denoting the propagation time required to send a bit from one site to another, limited by the speed of light. This parameter is dependent only upon the distance traversed and is significant when satellite links are used.
- *Transmission delay,* which is defined as the time required to transmit a block of data end-to-end. This parameter is dependent upon the bit rate of the network and the processing delays in the intermediate nodes, including routing and buffering.
- *Network delay,* which is composed of the transit and transmission delay components.
- *Interface Delay,* which is defined as the delay incurred between the time a sender is ready to begin sending a block of data and the time that the network is ready to transmit the data. This parameter is important in connection-oriented networks such as X.25 networks in which an end-to-end circuit must be established prior to transmission of data. Another situation in which this parameter comes into play is in token-ring LANs when transmission cannot begin until a free token has arrived.

The components of end-to-end delay are shown in Figure 2–2.

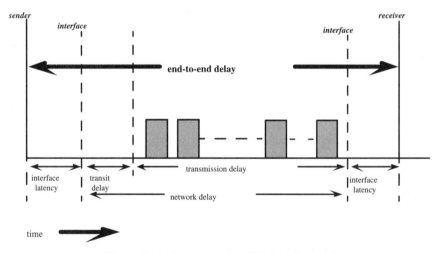

Figure 2–2 Components of End-to-End Delay

2.3.4 Round-trip Delay

For connection-oriented networks, when end-to-end acknowledgments are required, round-trip delay is useful. Round-trip delay is defined as the total time required for a sender to send a block of data through a network and receive an acknowledgment that the block was received correctly. Round-trip delay also plays a role in TCP networks running on top of connectionless IP networks.

When networks are very congested, this parameter sometimes gives a better picture of network performance than does end-to-end delay.

2.3.5 Delay Variation or Jitter

In digital video transmission, the video data stream and the audio data stream are often sent separately. In packet networks, these streams are further divided into discrete blocks of data, and each block is transmitted in sequence. If the network is able to transmit all of the blocks with a uniform latency, then each block would arrive at the destination after a uniform delay. Most of today's networks cannot guarantee a uniform delay to their users. Variations in delay are common. Whether the variations in delay are due to system imperfections in the network, either hardware or software, or due to traffic conditions within the network, the variations are commonly referred to as *jitter*. In designing a multimedia network, it is important to place an upper limit on the permissible jitter.

2.4 Characteristics of Multimedia Traffic Sources

Multimedia traffic often consists of long streams of data generated from digital video or audio sources. Even if these streams are broken up into packets or frames for network transport, it is important to observe the integrity of the streams themselves, and this in turn places constraints on the network performance parameters.

How do the network performance parameters affect multimedia traffic in a network? To answer this question, we must first examine the basic characteristics of multimedia traffic itself. Multimedia traffic is composed of five categories: audio, video, data, bit-mapped images, and graphics. Audio and video sources are mostly continuous in nature. The others are usually discrete, as defined in Chapter 1.

Multimedia data streams can be characterized according to the following: *throughput variation with time, time dependence, and bidirectional symmetry.* Let us examine these concepts briefly.

2.4.1 Throughput Variation with Time

Multimedia traffic can be characterized as *constant bit rate (CBR)* or *variable bit rate (VBR)*.

Constant Bit-Rate Traffic

Many multimedia applications, such as CD-ROM applications generate output at a *constant bit rate (CBR)*. For real-time applications involving CBR data streams, it is important for the network to transport these data streams at a constant bit rate also. Not to do so would require extensive buffering at each end-system. It is important to note here that many networks such as ISDN are by their very nature, CBR data transports.

Variable Bit Rate Traffic

Variable bit rate traffic has a data rate which varies with time. Such traffic often occurs in bursts or spurts. Such *bursty* traffic is characterized by random periods of relative inactivity interspersed with bursts of data. A bursty traffic source generates varying amounts of data at different time periods. A good measure of burstiness is given by the ratio of peak traffic rate over mean traffic rate over a given period of time.

Recent advances in compressed video technology have given rise to variable bit-rate traffic streams. VBR video streams arise from two primary factors: compression algorithms and scene contents. In a slow-moving scene, there is little need to retransmit, from frame to frame, the static parts of the scene, such as the background. However, in a motion video scene, the compression algorithm will generate, in each frame, new data representing the motion of the objects. VBR can either be used to conserve transmission capacity or to control display quality, as explained in chapter 4, VBR video streams are inherently bursty but can be adapted to CBR data networks. VBR traffic is relatively new in multimedia communications, so there are currently few standard ways to deal with such traffic.

2.4.2 Time Dependency

In applications such as video conferencing, the traffic generated is in real time, so that the end-to-end latency must be kept very low. For example, in video conferencing, experience has shown that the delay must be at most 150 ms in order for the participants to be unaware of its effects. In other applications, such as multimedia e-mail, the traffic generated is not required to be real time.

2.4.3 Bidirectional Symmetry

When two end-systems are connected by a network, the traffic over this connection is often asymmetric in nature. That is, in a bidirectional channel, the traffic in one direction might be significantly greater than the traffic going in the other direction. In many instances the forward channel is designed to carry stream traffic, while the reverse channel carries only short bursty traffic. Such a situation occurs in a cable network serving video-on-demand applications. The requester makes a selection on the reverse (control) channel, and the video data is sent to the requester on the forward data channel. In contrast to this, traffic generated in a peer-to-peer teleconference can generally be regarded as symmetric if all participants take an active part.

2.5 Factors That Affect Network Performance

We have defined the important network performance parameters to be throughput, error rate, delay, and delay jitter; now we examine the factors that degrade network performance with respect to these parameters.

2.5.1 Throughput Performance Factors

The throughput of most networks, whether local area or wide-area, varies with time. Sometimes the throughput can change very quickly because of sudden failures in network nodes or lines or because of traffic congestion when large streams of data are introduced into the network. Factors that affect overall throughput performance are:

- Node or link failures
- Congestion
- Bottlenecks
- Buffer capacity
- Flow control

Let us briefly examine each of these factors.

Node or Link Failures

Network nodes or transmission links whose operations are interrupted for whatever reason can cause congestion in other nodes and links in the immediate vicinity of the affected node (link). Such failures can lead to packet delays or loss, file transfer errors, and in some cases, total loss of connectivity. Even if the failure rates of network nodes or links are usually low, failures do occur, and measures must be taken to guard against such incidences.

Network Congestion

When a network is heavily loaded, the congestion may be due to heavy traffic or to bottlenecks. The capacity of a network is usually designed to accommodate average traffic demands. At certain times of the day or in emergency situations, the demand for network capacity exceeds the availability. In such times, the throughput of the network decreases with increasing load because of several factors:

- Many datagram networks start to drop packets when node buffers overflow.
- Network management procedures take effect to decrease traffic on certain links.
- Heavily loaded nodes become bottlenecks (see below).

Bottlenecks

Another reason for decline in throughput is the presence of *bottlenecks*. These may be due to node or link failures or due to inadequate link or node capacity. A case in point is the TransAtlantic satellite links that connect data networks in North America to those in Europe. Many of these satellite links have a throughput of 128 kbps. When these links connect two high-speed networks (such as T-1 or E-1) on opposite sides of the Atlantic, they represent a significant bottleneck. Users of the Internet who are trying to span the Atlantic or Pacific Oceans experience this kind of bottleneck often.

Buffer Capacity

For each end-to-end connection, there is a limited amount of buffer memory at the end-systems and at the network interfaces (Figure 2–3). Data is temporarily stored in these buffers when sending to or receiving from the network. In the transmission of large files, such as video frames, buffer capacity is very often inadequate to send or receive in real time.

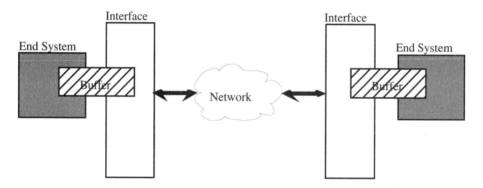

Figure 2–3 Buffering in End-to-End Connections

Flow Control

When buffer capacity at either end is a problem, *flow control protocols* are often invoked. Flow control is an end-to-end protocol that places limits on the rate of data transmission between two end-systems connected through a network. It is especially important if the receiving end-system does not have sufficient buffer capacity to accommodate all of the data that the sender wishes to transmit. The protocol is then invoked to *limit* or *meter* the data rate from the sender to prevent loss of data at the receiving end-system. When flow control is in operation, the end-to-end throughput is affected. Strictly speaking, flow control should not be classified as a network performance factor since it arises out of end-system buffer limitations.

2.5.2   Issues in Network Error Performance

Errors are a major concern in packet-switching networks. Network errors arise when:

- Individual bits in packets are inverted or lost
- Packets are lost in transit
- Packets are dropped or delayed
- Packets arrive out-of-order

Note that missing packets are either lost in transit (an inadvertent error) or dropped by an intermediate node (a deliberate error).

The error performance of these networks depends on the communications protocols employed. Some networks are *connection oriented*, in that before the packet or file is sent, an end-to-end connection must be first established. Other networks employ *connectionless* protocols, which is a *best-effort* strategy, in which packets are simply sent to the network without first establishing an end-to-end connection. Connectionless schemes work well for short messages, and connection-oriented approaches are best for stream traffic.

Individual Bit Errors

With the quality of today's data transmission networks, in particular, fiber optics networks, bit errors are very rare. Nonetheless, because of noise in the lines or packet switches, bit errors do sometimes occur. When that happens, error detecting codes employed in most packet switches are able to detect the presence of a bit error in the packet and can request retransmission of the faulty packet. Retransmission is either handled in intermediate nodes or handled on an end-to-end basis.

Packet Loss

In a connection-oriented network, when packets have bit errors or are lost or dropped, the receiving end-system is usually able to detect such a situation and inform the sending side of the problem. The receiving end-system does not always have precise information about which packets have errors or which packets have been lost or dropped. A standard approach to the problem is for the sender to retransmit the most recent packet(s) to the receiving end-system. In a network where forward error correction (FEC) is employed, the receiving end-system can often detect and correct packet bit errors without resorting to retransmission. However, FEC is not effective when entire packets are lost or dropped in transmission.

In the case of connectionless networks, packet loss or dropped packets are difficult, if not impossible, to detect. The primary reason for packets being dropped or lost in high-speed networks is insufficient buffer space at the receiving end-system due to congestion in the network. How to deal with this problem is discussed in later chapters.

Out-of-Order Packets

When a long file or stream of data is transmitted, usually the individual packets in the stream are numbered in sequence. It is the job of the receiving end-system to arrange the received packets in the numerical sequence in which they were originally sent. If the receiving end-system does not have such a capability, then the packets received could be out of sequence and thus might cause an error at the receiving end. Sometimes, an individual packet in a sequence might be lost, dropped, or greatly delayed. In such cases, the receiving end is not able to reconstruct the original packet sequence, and this situation might cause the receiving end-system to request a retransmission of either a portion or the entire packet sequence.

2.5.3 Network Delay Performance Issues

Some network delay is inevitable, as immutable as the laws of physics. If two end-systems are communicating via a satellite connection, then the one-way transit delay is approximately 0.25 seconds. Since the satellite is 36,000 kilometers above earth, the up-down time at the speed of light is about a quarter second. Other delays are due to the bit rate of the link: the broader the link bandwidth, the less the delay. Certain delays are not predictable. Congestion in a network can cause delays. Transmission errors can lead to delays. Physical problems in lines and switching nodes can also cause delays. These are classified as *random* delays and are the most troublesome to deal with.

The use of buffers at each end of the transmission can often smooth out delay problems. For example, in a long video stream, if the packets in the stream are received after undergoing varying delays, there would be much less jitter in the playback stream if it was buffered before playback. However, if the capacity of the buffers is inadequate and the video streams are very long, then jitter problems cannot be resolved.

It is very desirable for the network to present a constant, nonvarying delay to the end-systems. With a constant delay, i.e., with zero jitter, buffer resources could be allocated in advance, and the quality of received video and audio streams can be much higher

2.6 Multimedia Traffic Requirements for Networks

In this section, we review some of the principal requirements that multimedia traffic places on networks. These requirements are expressed in terms of the network performance characteristics of throughput, reliability (error), and latency. Other requirements such as multicast communications are also discussed. Note that the requirements are only introduced here. Specific, detailed requirements for video and audio traffic, with or without compression, are discussed in later chapters.

2.6.1 Throughput Requirements

The multimedia requirements for throughput performance are discussed below.

High Transmission Bandwidth Requirement

Since multimedia traffic consists of more than bursty data and often includes real-time video and audio traffic, it is important that networks carrying multimedia information have enough transmission bandwidth available to handle this traffic. This requirement also means that networks must have the capacity to handle multiple sources of such information. In periods of congestion, insufficient available bandwidth could often lead to longer end-to-end delays, as well as packet loss.

High Storage Bandwidth Requirement

In high-throughput networks, it is important that the receiving end-system have sufficient buffer capacity to receive the incoming multimedia traffic. In addition, it is necessary that the buffer's input data rate be high enough to accommodate the incoming data stream from the network. This input data rate is sometimes referred to as the buffer's storage bandwidth. While not strictly a network requirement, storage bandwidth is relevant enough to be included in this discussion.

Streaming Requirement

A multimedia network must be able to handle long streams of traffic, such as that coming from video and/or audio sources. What this boils down to is that the network must have sufficient throughput capacity to ensure the availability of high bandwidth channels for extended periods of time. For example, it is not sufficient for a network to offer a user a 5-second time-slot at 1.5 Mbps if the user needs to send a stream of traffic of 30 megabits. However, the network would meet the streaming requirement if it can offer the *continuous* availability of a 1.5 Mbps channel to the user. If there are many streams on the net at any one time, the network must have available throughput capacity equal to or greater than the *aggregate* bit rate of the streams.

2.6.2 Reliability (Error Control) Requirements

It is difficult to precisely quantify error control requirements for multimedia networks because multimedia applications are, to a certain extent, tolerant of transmission network errors. Part of the reason for this tolerance is traceable to the limits of human sensory perception. If, for example, an error occurs in a single packet in a long video stream, quite often the error is undetectable by the human eye. A similar situation occurs in audio transmission. However, the visual and auditory senses in a human are not equally tolerant of such errors. Experience has shown that dropped packets are more noticeable in an audio stream than in a video stream. Similarly, a dropped packet in a text stream would be much more noticeable than in either video or audio streams.

Error control requirements are also difficult to quantify because in many cases the requirements for error control and for end-to-end latency are contradictory. Contradictions arise because many error control schemes involve detection and retransmission of the packet in error or lost. In some instances, retransmission must be carried out on an end-to-end basis, which significantly increases delay. For real-time video or audio transmission, delay is a more important performance issue than error rate, so in many cases, it is preferable to forget the error and simply work with the received data stream as is.

In later chapters we review specific situations in which error control is an important issue.

2.6.3 Delay Requirements

Multimedia data often takes the form of multiple streams of data, such as video and audio streams, which constitute different but closely interrelated parts of video scenes. In real-time applications, the video and audio streams must be transmitted through the network with minimum delay and should arrive at the same time. However, with the help of buffering, simultaneous arrival of the parallel audio and video streams is not absolutely necessary. When both streams do arrive at essentially the same time, we say that they are *synchronous* streams.

There are variations of the synchronicity concept which should be discussed. The following definitions are instructive.

Asynchronous. A network connection is said to be asynchronous if there is no upper limit on the delay through the network. This definition implies that the latency can take on any value.

Synchronous. The term means *occurring at the same time or at the same rate* with a regular or predictable temporal relationship. An end-to-end connection is synchronous if two data streams traverse the network at essentially the same rate and arrive at the destination end-system at the same time. Since network delay cannot be less than transit delay because of the speed of light limitation, a synchronous network imposes only a fixed, predictable delay over and above the transit delay. If the additional delay imposed by the network is variable, then the following definition is relevant.

Isochronous. A network connection is isochronous if there is both an upper bound and a lower bound to the latency and the difference between upper and lower bounds is small. If the lower bound is some value T, then the upper bound is T + dT, where dT is small. The value dT can be regarded as delay variation or delay jitter, which must be kept at a minimum.

It is difficult to specify precisely what the terms small and minimum delays are because they really depend upon the requirements of the underlying application. In real-time applications, the end-to-end latency must be kept small, but the delay variations should also be kept at a minimum. The problem of delay variations is also important when dealing with multiparty applications. Interesting empirical results on this topic can be found in Steinmetz [3].

2.7 Quality of Service

Quality of Service (QoS) is a term indicating how well a network performs in dealing with a multimedia application. Since different applications impose various performance requirements upon the network, the individual applications have differing expectations of how well the network carries out its tasks. These expectations are expressed in terms of QoS parameters.

There are many ways to express QoS requirements. QoS parameters can be expressed as maximum allowable delay, delay jitter, throughput, error rates, etc. Certain applications, such as real-time conferencing, might impose QoS requirements on latency and throughput. Others might require zero error rates but not have tight restrictions on latency or throughput.

Since QoS parameters can be defined explicitly, they can form a basis to determine whether a network is able to meet the QoS requirements for a given application. For example, if a multiparty video conference requires high bandwidth and low delay on a sustained basis over several hours, and if the network used is subject to congestion over that time period, the network might not be able to offer the quality of service demanded of it for the application.

In recent experimental multimedia systems, new QoS concepts have emerged that may be implemented as standard services in future networks. Let us next examine some of these concepts.

Resource Reservation and Scheduling

If an application "knows" in advance that it requires certain resources from the network, say, a given bandwidth allocation, it can make a reservation with the network for those resources for the period in question. The network can either deny the request or schedule the application for that period and reserve the resources requested.

Resource Negotiations

If the network administrator feels that the requested resources might overtax the capabilities of the network, it can negotiate with the requester and offer lower QoS parameters. A mutually acceptable set of QoS parameters can then be negotiated.

Admission Control

If the QoS demands of the particular application are so high that the network cannot meet them, the network has the choice of not letting the application on to the network. This feature is known as admission control.

Guaranteed QoS

Suppose the network wants to attract a particular user to its service: It can offer the potential user a guaranteed quality of service, in terms of available bandwidth, upper limits on end-to-end delay, maximum allowable error rates, etc.

This offer means that the user will expect a guaranteed level of service from the network. Whether these guarantees are statistical or absolute depends upon the negotiations between the user and the network.

At this time, there are few, if any, data networks that offer hard QoS guarantees. A network can point to the performance of its service on a statistical basis, over a given period of time. With information about the underlying technology employed in the network and with operational experience, an accurate estimate of a network's average throughput, error rates, and latency characteristics can be obtained. So, a network is able to determine a priori if it could generally meet the QoS requests of a particular application. More details on QoS are given in later chapters.

References

[1]   S. A. Bly, S. R. Harrison, and S. Irwin, "Media Spaces: Bringing People Together in a Video, Audio, and Computing Environment," *Communications of the ACM,* 36(1), Jan. 1993

[2]   M. Moran and R. Gusella, "System Support for Efficient Dynamically Configurable Multi-Party Interactive Multimedia Applications," *Proceedings of the Third International Workshop on Network and Operating System Support for Digital and Audio,* San Diego, November 1992.

[3]   R. Steinmetz, "Human Perception of Jitter and Media Synchronization," *IEEE JSAC* 14(1):61-72, 1996

[4]   C. Szyperski and G. Ventre, "Efficient Support for Multimedia Communications," in *Multimedia Transport and Teleservices,* edited by D. Hutchison et al. Springer-Verlag, 1994:185-198.

[5]   S. L. Teger, "Multimedia — From Vision to Reality," *AT&T Technical Journal,*. 74(5):4–13, September/October 1995.

Compression Methods

Ralf Keller

In this chapter, we discuss compression methods for digital audio and video streams. We first present an introduction to compression techniques, followed by a description of basic coding methods. Afterward, we look into the details of some well-known standardized and non-standardized hybrid coding techniques for digital audio and video streams.

3.1 Introduction to Compression Methods

With today's technology, only compression makes the storage and transmission of digital audio and video streams possible. Either the data rates of the used storage and transmission media are too low for uncompressed digital audio and video data or the requested prices for the data rates are too high. Often both reasons are true at the same time.

Data compression, as the term is used in this book, can be defined as a means of reducing the size of blocks of data by removing unused and redundant material. An example of unused material is a silence period in a telephone call, which is relatively easy to detect and suppress. In contrast, it is often very difficult to extract redundant material from digital audio and video streams, i.e., to define which fraction of the information content can be eliminated without loss of essential information. Simple forms of redundancy like data replication are easy to describe and can be coded efficiently, but much more redundancy can be exploited for compression by an in-depth look at the human perceptive faculty.

Several types of redundancy, as shown in the remainder of this chapter, can be exploited within compression methods:

- *Spatial redundancy.* The values of neighboring pixels are strongly correlated in almost all natural images.
- *Redundancy in scale.* Important image features such as straight edges and constant regions are invariant under rescaling.
- *Redundancy in frequency.* In images composed of more than one spectral band, the spectral values for the same pixel location are often correlated and an audio signal can completely mask a sufficiently weaker signal in its frequency-vicinity.
- *Temporal redundancy.* Adjacent frames in a video sequence often show very little change, and a strong audio signal in a given time block can mask an adequately lower distortion in a previous or future block.
- ☞ *Stereo redundancy.* Audio coding methods can take advantage of the correlations between stereo channels.

Spatial redundancy and redundancy in scale are typical for video streams, whereas stereo redundancy can be exploited only in the coding of audio streams. All other types of redundancy can be found in audio as well as in video streams.

Most of the simpler compression methods described in the next section exploit only one kind of redundancy. Spatial redundancies and redundancies in frequency are often removed by transform coding. Temporal redundancy is exploited by techniques that encode only the difference between adjacent samples or frames, such as differential pulse code modulation [32] and motion compensation [7]. More sophisticated compression methods, as described in the remainder of this chapter, are combinations of simpler compression methods, thus exploiting several types of redundancy to gain better compression.

Compression methods can be classified by several characteristics, as depicted in Table 3–1 (derived from Milde [37]), together with a brief description. A certain compression method can often be classified by more than one of these characteristics, for example, a lossless, intraframe, symmetrical compression method.

Many of the first compression methods were developed for the lossless compression of data, e.g., Huffman coding [18] and Ziv-Lempel coding [57]. But it soon became clear that lossless compression methods applied to digital audio and video are inadequate for transmission media with low bandwidth (e.g., ISDN) or for devices with low data throughput (e.g., CD-ROM). Therefore, lossy compression methods have been developed to reduce the data rates even further. Of main interest are techniques that attempt to extract out of the data stream the information that is barely noticeable to a human user, thereby increasing the compression factor [30].

Characteristics	Description
lossless	Original data can be recovered precisely.
lossy	Not lossless.
intraframe	Frames are coded independently.
interframe	Frames are coded with references to previous and/or future frames, i.e., temporal redundancies between frames are taken into account.
symmetrical	Encoding and decoding time are almost equal.
asymmetrical	Coding time considerably exceeds decoding time.
real-time	Encoding-decoding delay should not exceed 50 ms.
scalable	Frames are coded in different resolutions or quality levels.

Table 3–1   Characteristics of Compression Methods

The relation between the perceptible quality of an audio or video data stream and the bandwidth requirements is depicted in Figure 3–1.

By application of lossless compression methods, the bandwidth requirement can be reduced to a certain extent without affecting the quality. To reduce the bandwidth requirement even further, lossy compression techniques must be used, i.e., bandwidth reduced below a certain limit results in lower quality. Within recent years, new compression methods have been developed that can reduce the bandwidth requirements enormously without having to reduce quality to the same

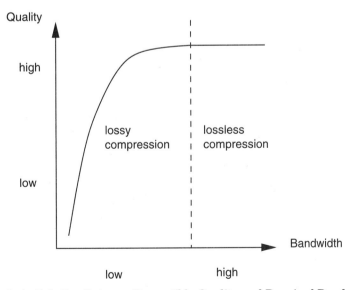

Figure 3–1   Relation Between Perceptible Quality and Required Bandwidth

extent, i.e., only information that is not or hardly perceptible to a human user is removed from the data stream.

For their use in multimedia systems, we can distinguish between *entropy*, *source*, and *hybrid coding techniques* (see also Steinmetz [47] and Table 3–2).

Coding Technique	Examples
Entropy coding	Arithmetic coding Huffman coding Run-length coding
Source coding	Differential pulse code modulation Discrete cosine transform Discrete wavelet transform Fourier transform Iterated function system Motion-compensated prediction
Hybrid coding	Fractal image compression H.261 H.263 JPEG MPEG video MPEG audio Perceptual Audio Coder Wavelet image compression

Table 3–2   Classification of Coding Techniques for Multimedia Systems

Entropy is defined as the average information content of given data. It defines the minimum number of bits needed to represent the information content without information loss. Entropy coding tries to come as close as possible to this theoretical lower limit. The decompression process reconstructs the original data completely; therefore, entropy encoding is a lossless technique. The data stream to be compressed is considered as a simple digital sequence, and the semantics of the data are ignored, i.e., entropy encoding is used for media regardless of their specific characteristics.

Source coding processes original data such that a distinction between relevant and irrelevant data is possible. It takes into account the semantics of the data. Removal of the irrelevant data compresses the original data stream. In contrast to entropy coding, source coding is often a lossy process. In the case of a lossy compression technique, the original data stream and the decoded data stream are similar but not identical.

Hybrid coding uses a combination of entropy coding and source coding techniques. Often, several different entropy coding and source coding techniques are grouped together to form a new hybrid coding technique (see Figure 3–2). As a rule, the output data stream produced by the source coding steps is used as the input data stream for the entropy coding steps.

The preparation step in Figure 3–2 includes the analog-to-digital conversion, generating an appropriate digital representation of the information. A brief introduction to this topic can be found in Chapter 1. More information about the preparation step is available in the literature, e.g., [53][54].

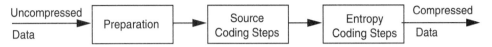

Figure 3–2 Major Encoding Steps of Hybrid Coding Techniques

In the following sections, we first present a brief description of basic coding methods as used by different hybrid coding methods. Readers already familiar with entropy and source coding methods can skip this section. Then, we go into the details of well-known standardized and nonstandardized hybrid coding techniques for digital audio and video streams. Inasmuch as printed media (like books) cannot contain examples of continuous media (like audio and video streams), we present only still image examples to illustrate certain characteristics of compression methods.

3.2 Basic Coding Methods

Entropy coding methods, as already mentioned, use no semantics of the data; the data stream to be processed is considered as a bit or byte stream. Three examples of entropy coding methods are presented in the following subsections: *run-length coding*, *Huffman coding*, and *arithmetic coding*.

3.2.1 Run-Length Coding

Most sampled image, audio, and video data can be compressed by suppressing sequences of same bytes. These sequences are replaced by the number of occurrences and the repetitive byte pattern itself. Obviously, the attainable compression factor depends on the input data. Using an exclamation mark as special flag to indicate run-length coding, the following example shows how the data in the example can be compressed by replacing the sequence of six characters "N" with "!6N":

    Uncompressed data:UNNNNNNIMANNHEIM
    Run-length coded: U!6NIMANNHEIM

It should be clear that it is useless to replace sequences of characters shorter than four by this run-length code because no compression can be achieved. For example, replacing the sequence of two characters "N" with run-length code

"!2N" would increase the length of the code by one byte. If the special flag in our example occurs in the data, it has to be replaced by two exclamation marks ("byte stuffing").

The simple algorithm presented above is easily optimized; e.g., instead of sequences of single characters, longer sequences of different characters can also be replaced. This extension requires that either the length of the sequence has to be coded or a sequence end flag has to be used. Many variants of run-length coding exist; see Steinmetz [47] for an overview.

Different characters do not have to be encoded with a fixed number of bits. A well-known example is the Morse alphabet: frequently occurring characters are encoded with strings shorter than those used for characters seldom occurring. If the probability of the occurrence of each character is given, characters can be encoded with different number of bits. There are different techniques based on these methods, the two most prominent of which are discussed below.

3.2.2 Huffman Coding

Huffman [18] developed a compression method to determine the optimal code for given data, i.e., using the minimum number of bits given the probability. Hence, the length in bits of the coded characters will differ. The shortest code is assigned to those characters that occur most frequently.

A binary tree illustrates the process of determining a Huffman code. The characters to be encoded are placed in the leaves of the tree. Every node is labeled with the probability p of the occurrence of one of the characters in the subtree. The tree is constructed by successively combining the two nodes with lowest probability until the root is reached. Obviously, the root of the complete tree is labeled with probability 1. After the construction of the tree, all edges are assigned a value of 0 and 1. The Huffman code of each input character is the sequences of labels on the way from the root to the leave.

The following example illustrates this algorithm for a binary code. A detailed description of the general algorithm can be found, e.g., in Heise [15]. In Figure 3–3, the characters A, B, C, and D have the following probability of occurrence:

$$p(A) = 3/4, p(B) = 1/8, p(C) = p(D) = 1/16$$

Figure 3–3 depicts the binary tree together with the reduction process. The result of the Huffman coding is the following code that is stored in a so-called *Huffman table*:

$$w(A) = 1, w(B) = 01, w(C) = 001, w(D) = 000$$

The most expensive operation in determining a Huffman code is the addition of floats, more specifically, the addition of the probability of occurrence in the reduction process. In contrast, the decoder has to perform only a simple (and fast) table lookup. Therefore, the decoder needs the Huffman table used by the coder. In dependence on the implementation, this table is either part of the data stream, already known by the decoder, or has to be calculated at runtime.

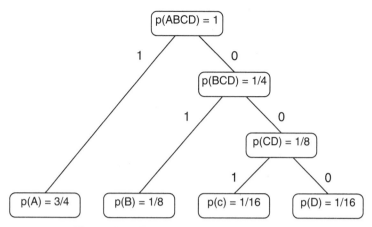

Figure 3–3 Example of Huffman Coding

Many variants of Huffman coding have been developed to accelerate the coding and decoding process, to optimize the resulting Huffman code, or to enable an efficient implementation. In audio and video coding methods, standard Huffman tables are often used, i.e., the tables are known in advance to both encoder and decoder. The advantage is faster encoding because the tables need not be calculated at coding time. A disadvantage of standard Huffman tables is the slightly worse compression factor because the tables are not necessarily optimal for the data to be encoded. Therefore, compression methods to be performed in real time often use standard Huffman tables because of faster encoding. If highest quality is required but encoding time is not that important, optimal Huffman tables should be used.

Often, not all input characters have a code representation in the Huffman table, rather only those characters with a high probability of occurrence. All others are coded directly and marked with a special flag. This technique is useful when the number of different input characters is very high but only a few of them have a high probability of occurrence.

3.2.3 Arithmetic Coding

In theory, arithmetic coding [17] is as optimal as Huffman coding, i.e., the length of the encoded data is minimal. In practice, arithmetic coding often produces slightly better results in audio and video coding because it works with floats instead of the characters used in Huffman coding, thereby enabling a closer approximation of the mathematical optimum. But the use of floats is also computationally more expensive and complicates an implementation, e.g., the binary representation of floats with very low values poses many problems. In addition, the algorithm for arithmetic coding is covered by patents held by IBM, AT&T, and Mitsubishi.

Practical experiments have shown that the average compression achieved by arithmetic and Huffman coding is quite similar, but also that Huffman coding is faster than arithmetic coding. Therefore, many hybrid compression methods use an optimized Huffman coding as one of the entropy coding steps

The entropy coding methods described above compress data by removing redundant information in a lossless process. If image, audio, and video are to be compressed even further, the human perceptive faculty has to be considered. Source coding divides the original data into relevant and irrelevant information, thereby enabling consecutive processing steps to remove the irrelevant data. In contrast to entropy coding, source coding can be lossy.

Which information is considered relevant and which irrelevant depends on the source coding method. Much more importantly, different source coding techniques exploit different characteristics of the human perceptive faculty. One method to separate relevant from irrelevant information is transform encoding, which transforms data into a different mathematical model better suited for the purpose of separation.

The most widely known transform coding is *discrete cosine transform* (DCT). Other examples are *Fourier transform* (FT) and *discrete wavelet transform* (DWT). For all transform encodings, an inverse function must exist to enable reconstruction of the relevant information by the decoder. In addition, rapid calculation is an important requirement. In the following section, discrete cosine transform is presented and a comparison between DCT and FT is outlined. More details on DWT are presented in "Wavelet Image Compression" on page 62.

3.2.4 Discrete Cosine Transform

All these above-mentioned requirements are fulfilled by a one-dimensional or two-dimensional discrete cosine transform. A one-dimensional DCT is used in audio compression methods; the only dimension of interest is time. In image compression methods, however, the vertical and horizontal dimensions have to be considered, thus requiring a two-dimensional DCT.

The concrete formulas of forward and inverse DCT are of no interest for a further understanding and can be found in the literature, e.g., [40][44][48]. To explain how DCT works, we now discuss image coding, using a two-dimensional DCT. An image is subdivided into 8×8 blocks of samples. Each of these 8×8 blocks of samples of the original image is mapped to the frequency domain, i.e., it is represented as a composition of DCT *basis functions* with appropriately chosen 64 coefficients, representing different horizontal and vertical intensities. The DCT basis functions are shown in Figure 3–4.

The human eye is highly sensitive at low-intensity levels, whereas its sensitivity is greatly reduced at high-intensity levels. A straightforward conclusion: A reduction of the number of high-frequency DCT coefficients weakly affects image quality. In the frequency domain, it is very easy to separate the high-intensity values from the low-intensity levels because their position in the 8×8 block of

0 1 2 3 4 5 6 7

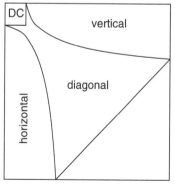

increasing vertical frequency

rizontal frequency

DCT Basis Functions

expressed as a linear combination of these 64

shows the frequency distribution and block

ordered in the zigzag sequence of Figure 3–6,
hly with increasing frequency and decreasing
wer frequency (typically, with higher values)
ncoding of higher frequencies (with typically
oefficient at the first position in the zigzag

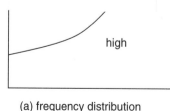

(a) frequency distribution

DC

vertical

diagonal

horizontal

high

(b) block features

Figure 3–5 Frequency Distribution
(a) of two-dimensional DCT coefficients and block features (b) they represent

sequence is called the *DC coefficient*, all others *AC coefficients*. The DC coefficient has zero frequency in both the horizontal and vertical directions and represents the average color of the 8×8 block of samples (see also Figure 3–4).

In theory, up to this point in encoding, no information has been lost, i.e., the original 8×8 block of samples can be reconstructed. In practice, information is lost because of the limited precision of calculations. Therefore, DCT is used in many applications only with integer numbers and with optimized calculations. Fast implementations of forward DCT need less than 1 multiplication and 9 additions on the average to calculate a DCT coefficient, as opposed to the 64 multiplications and 63 additions required by the original formula [19].

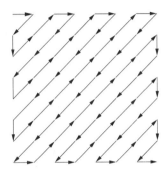

Figure 3–6 Zigzag Sequence: Order of AC Coefficients with Increasing Frequency

Transform encodings like DCT do not compress the data. Compression is achieved by source coding steps after a so-called *quantization*. The quantization is usually performed by dividing each DCT coefficient by an integer and by rounding the result to an integer. Careful choice of quantization values allows most DCT coefficients with higher frequencies to be quantized to a zero value. These quantized DCT coefficients can now be compressed very effectively by first applying a variant of run-length coding, thereby replacing sequences of zeros by a run-length code, and then using a modified Huffman coding.

A drawback of using blocks of samples in coding methods as described above is the loss of correlation between block boundaries. At low bit-rates, i.e., at high compression factors, visual artifacts such as block boundaries tend to appear [19]. The example in Figure 3–7 shows both the original and coded/decoded image, using a DCT-based coding method with a high compression factor (1:52).

As is shown in the remainder of this chapter, discrete cosine transform is used in many standardized image, audio, and video compression methods. It has shown its superiority in reduction of the redundancy of a wide range of signals such as speech, TV signals, color print images, infrared images, etc. Fourier transform, another transform encoding, is at the forefront of digital signal processing because it decomposes a signal into its frequency components and has simple mathematical properties [44]. For example, many algorithms developed within the MoCa project are based on the Fourier transform. The MoCa project

Original Lena image　　　　　Lena image after coding/decoding

Figure 3–7　Example of blocking artifacts using DCT-based coding methods

at the University of Mannheim aims at the automatic analysis of audio and video streams [12][42].

Of the three above-mentioned transform encoding methods, only discrete wavelet transform has not yet been described in this chapter. Since we focus on DWT only with respect to image compression, DWT is introduced in the corresponding section (3.3.8)

Instead of compressing single bytes or sequences of bytes as in transform coding, one can use a differential coding. These techniques can exploit the temporal redundancy of digital audio and video signals. Differential pulse code modulation and motion compensation, two important representatives of this class of coding methods, are presented in the following two sections.

3.2.5　Differential Pulse Code Modulation

One of the simplest source coding methods is differential pulse code modulation (DPCM [32]). DPCM reduces the value range of numerical input characters in such a way that successive entropy coding methods might achieve better results. DPCM is often applied in audio techniques. A sequence of PCM-coded samples can be sufficiently represented even if only the first PCM-coded sample is stored as a whole and all following samples are stored as a difference to the previous one, i.e., they are stored as predicted values.

Take, for example, the sequence 10, 12, 14, 16, 18, 20. A run-length coding method can achieve no compression, since all numbers are different. If DPCM coding is applied first, generating, for example, as the new sequence 12, 2, 2, 2, 2, 2, run-length coding can in turn compress the new sequence significantly, e.g., to 12!52 (see section 3.2.1).

DPCM, as described above, is a lossless coding method and has two disadvantages: First, since only a very small number of bits (e.g., 1 bit) are used to code the difference, the quantization error increases over time. Second, the coding of audio signals with very strong signal oscillation causes high quantization errors. Lossy modifications of DPCM have been developed to overcome these disadvantages (see section 3.4.1 for an example).

3.2.6 Motion-Compensated Prediction

By motion-compensated prediction, temporal redundancies between two frames in a video sequence can be exploited. This technique is used in all interframe coding methods. Temporal redundancies can arise, e.g., from movements of objects in front of a stationary background and from camera movements.

The basic concept of motion-compensated prediction is to look for a certain area (block) in a previous or subsequent frame (reference frame) that matches very closely an area of the same size in the current frame. If successful, i.e., a best matching block is found, then the difference signal (DPCM-coded) between the block intensity values of the block in the current frame and the block in the reference frame is calculated. In addition, the motion vector, which represents the translation of the corresponding blocks in both x- and y-direction, is determined. Together, the difference signal and the motion vector represent the deviation between reference block and predicted block; therefore, they are also termed *prediction error*.

In general, three types of motion-compensated prediction can be distinguished:

- In *unidirectional motion-compensated prediction*, only a previous or a subsequent frame is used for prediction.
- In *bidirectional motion-compensated prediction*, a previous reference frame as well as a subsequent reference frame is used in determining the motion vector for each block. Only the motion vector corresponding to the previous or future reference frame that results in the smallest matching error is used. Bidirectional motion-compensated prediction deals with the problem of covered and uncovered areas [34].
- In *interpolative motion-compensated prediction*, the prediction of the previous reference frame and the prediction of the future reference frame are averaged. Obviously, two motion vectors have to be stored or transmitted, respectively.

An example of an unidirectional motion-compensated prediction with a previous reference frame is depicted in Figure 3–8 (derived from [7]). Since the prediction goes from the previous reference frame to the current frame, this type is also termed forward motion-compensated prediction.

Usually a search area, a fixed distance in x- and y-direction, can be defined to find a match for a block, thereby limiting the search range and reducing the costs of extracting motion information (the process of extracting motion information is

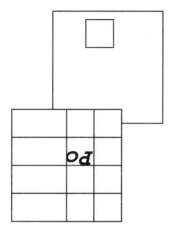

Figure 3–8 Forward Motion-Compensated Prediction

called motion estimation). The details of motion compensation techniques like search method and search area vary among different interframe coding methods. A good comparison of different motion-compensation techniques can be found in Chan et al.[7]

3.3 Video Compression

This section contains a description of the three hybrid coding methods: JPEG, H.261, and MPEG-1 Video, standardized by ISO and ITU. In addition, non-standardized but well-known coding methods and further developments of existing methods, i.e., H.263, MPEG-2 and MPEG-4, are outlined. Also wavelet and fractal image compression, both new approaches in the field of image compression, are introduced. Because the basic coding methods have been described in previous sections, only the differences between the coding methods are presented.

Though JPEG is an image compression method, it is also used as an intraframe coding method for digital video. This method is termed *Motion-JPEG (MJPEG)*. In Motion-JPEG, the JPEG method is applied to each frame of the video sequence to be coded.

To uniquely describe the hybrid coding methods, we subdivide the coding process into three steps: image preparation, image processing, and entropy coding.

3.3.1 Nonstandardized Techniques

The hybrid coding methods H.261, JPEG and MPEG Video are the present-day endpoints of a long development history. Various methods for the compression of digital video have been developed by research groups in industry and at universities. The predecessors of most of the other methods are *Block Truncation Coding* (BTC [11]) for the compression of monochrome images and *Color Cell Compression* (CCC [5]) for the compression of color images.

Block truncation coding and all its successors rely on the ability of the human visual system to perform spatial integration over the displayed image or frame. When viewing the displayed area from distance, our eyes average fine detail within the small area (blocks) and perceive the local average value of the signal. This phenomenon can be exploited by the following technique: For each block of pixels, only two color values and a binary pattern are stored; the binary pattern indicates which of the two colors is associated with that pixel. Some methods encode blocks that exhibit little variation with one color rather than use a binary pattern with two colors.

Many methods have been developed for the compression of sequences of digital images, extending BTC and CCC: *Extended Color Cell Compression* (XCCC [33]), developed at the University of Mannheim and, promoted by industry, the coding methods *Digital Video Interactive* (DVI [14]), *QuickTime* [1] and *Software Motion Pictures* (SMP [38]), to mention only a few. These methods differ in how they determine whether to represent a block with one or two colors and in how they estimate and represent the binary pattern and the two colors.

Common to all these compression methods is the use of color lookup tables to achieve data reduction, and implementation of a software-only decoder on standard PCs or low-end workstations. For example, Indeo from Intel is the software version of the hardware-based DVI-system. But the color-lookup technology and software-only decodability also imply some disadvantages in comparison to the standardized methods: both image quality and compression ratio are clearly lower [46]. Typical artifacts of CCC-based coding methods are depicted in Figure 3–9, showing the Lena image compressed to 1:10 of its original size.

The following section, which is dedicated to JPEG, introduces some basic compression concepts that can also be found in both H.261 and MPEG-1 Video.

3.3.2   JPEG

JPEG stands for *Joint Photographic Experts Group—Joint* because the development of this standard was a joint effort by ITU (formerly CCITT) and ISO [52]. JPEG is a compression standard for continuous-tone still images (gray scale and color) [20]. It is supposed to be a generic standard for many applications. The JPEG standard offers one lossless and three lossy encoding modes:

- Lossless mode
- Sequential DCT-based mode (baseline mode)
- Progressive DCT-based mode
- Hierarchical mode

Of these modes, only the *baseline mode* is presented in more detail because only this mode must be supported by each JPEG implementation. It is by far the most-implemented JPEG method, thereby providing a standard decoding

Figure 3–9 Typical Artifacts Caused by a CCC-based Coding Method

method. Many of the up-to-date JPEG coders use only the baseline method. The relationship between the four encoding modes is depicted in Figure 3–10.

The *baseline mode* achieves compression by first applying DCT, then quantizing, and, finally, entropy coding the corresponding DCT coefficients. The basic coding units are 8×8 pixel blocks into which the whole image is divided. Each image component is encoded in a single left-to-right, top-to-bottom scan. Figure 3–11 and Figure 3–12 depict the processing steps of a baseline mode encoder and decoder, respectively

In *progressive mode*, each image component is encoded in multiple scans rather than in a single scan. The first scan encodes a rough but recognizable version of the image, which is refined by successive scans until the requested image quality is reached. Thus, preview images, for example, can be generated, which

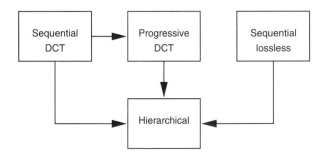

Figure 3–10 Four Coding Modes of JPEG

can be transmitted and presented very quickly in comparison to the total transmission time, without the need to completely decode the image.

The *lossless mode* enables a lossless image compression. Rather than transform encoding combined with quantization as source coding steps, a simple predictive method is used. The lossless mode cannot gain such high compression factors as the lossy JPEG modes deliver; nonetheless, many applications require lossless storage of images, e.g., X-ray photographs.

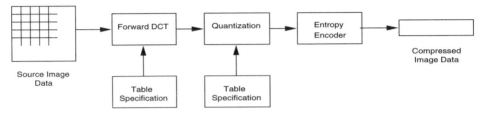

Figure 3–11 Processing Steps of Baseline Mode Encoder

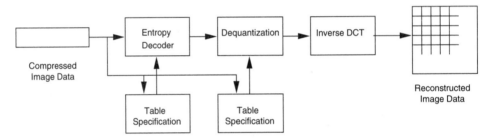

Figure 3–12 Processing Steps of Baseline Mode Decoder

The *hierarchical mode* encodes an image at multiple resolutions, differing by a factor of 2 in horizontal and/or vertical resolution. Lower-resolution versions can be accessed without the need to first decompress the image at its full resolution. The drawback to having multiple resolutions present within one file is clear: the compression ratio is lower in comparison than it would be were only one resolution present. The hierarchical mode uses algorithms defined in the other three modes, as depicted in Figure 3–10. See Pennebaker and Mitchell [40] for an in-depth description of all four modes.

Image Preparation in JPEG

JPEG introduces a very general image model, able to describe most of the well-known, two-dimensional image representations. Each source image must have a rectangular format and may consist of at least 1 and at most 255 components or planes. Each component may have a different number of pixels in the horizontal than in the vertical axis, but all pixels of all components must be coded with the same number of bits. Each mode defines its own precision. For example, the planes may be assigned to the three colors RGB (red, green, blue), or to the three components of YUV (one luminance and two chrominance components).

A gray-scale image will, in most cases, consist of a single component. An RGB color representation has three components of equal resolution. In YUV color representation, the chrominance components of an image are often down-sampled to reduce the amount of storage. For example, in a 4:2:2 format, the chrominance components have half the horizontal resolution of the luminance component; the vertical resolution is equal. In a 4:1:1 format, the chrominance components have half the horizontal and half the vertical resolution of the luminance component. This reduction of resolution is motivated by the fact that the human visual system is more sensitive to the luminance component and less sensitive to the chrominance components of visual light.

After the separation of the source image into color components, each component is further subdivided into data units of 8×8 pixel blocks. In most cases, the data units are processed component-by-component and passed, as shown in Figure 3–11, in this generated order, to the forward DCT for further processing. As shown in Figure 3–13, for one component, the order of processing of the data units is left-to-right and top-to-bottom, one component after the other; this is known as *noninterleaved* data ordering. When this noninterleaved mode is used, a JPEG decoder must first decode the image completely before the image can be displayed. Otherwise, the display would initially present the partially decoded image in wrong colors. This disadvantage can be overcome by interleaved processing of data units [40].

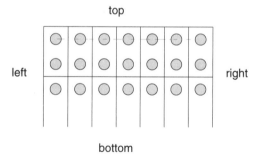

Figure 3–13 Noninterleaved Data Ordering

Picture Processing in JPEG

The baseline mode achieves compression by first applying a two-dimensional DCT, then quantizing, and, finally, entropy coding the corresponding DCT coefficients, as described in section 3.2.4. Corresponding to the 64 DCT coefficients, a 64-element quantization table is used. The DC coefficient and each AC coefficient are divided by the corresponding quantization table entry. Then, the values are rounded. For JPEG, there can be more than one quantization table specified by the application. Quantization tables can vary, depending on the characteristics of the intended display, the viewing distance, and the amount of noise in the source. The standard does not dictate quantization tables; it only makes some recommendations. Examples for quantization tables can be found in the literature [40], [41].

An application can influence the quantization by means of a *quantization factor* (Q factor). The quantization tables are multiplied by the Q factor. As the Q factor increases, the quality decreases in favor of the increased compression factor. If the Q factor is too large, visual artifacts in the form of block boundaries tend to appear because the correlation between block boundaries gets lost. An example for this kind of visual artifact is given in the right Lena image in Figure 3–7.

Entropy Encoding in JPEG

JPEG specifies both Huffman and arithmetic encoding as methods for entropy coding. However, for the baseline mode of JPEG discussed in this section, only Huffman encoding may be used. Usually, the inputs to the entropy encoder are a few non-zero and many zero-valued coefficients. The zigzag-reordered data stream of quantized DCT coefficients is processed in two steps. The first step is a run-length encoding of zero values of the quantized AC coefficients. Each non-zero AC coefficient is represented by a pair of symbols, of which the first symbol stores the number (run-length) of preceding zero values and the number of bits (size) needed to encode its non-zero amplitude. The second symbol represents the non-zero amplitude. The quantized DC coefficient is encoded as the difference to the DC coefficient of the previous block; only the size information in the first symbol is needed. The first symbol is entropy encoded using a Huffman table, and the second symbol is represented by a variable-length integer value.

As already mentioned, Huffman encoding is, in contrast to arithmetic coding, not protected by any patent. But also in contrast to arithmetic coding, the application must provide encoding tables since the JPEG standard does not predefine any of them (see also Figure 3–11). The baseline mode allows the use of different Huffman tables for AC and DC coefficients.

This section was dedicated to JPEG, a standardized intraframe hybrid coding method. In discussing H.261 in the next section, we present an example of interframe coding, namely, motion-compensated prediction. Instead of giving a complete description of H.261, we outline only the differences from JPEG.

3.3.3   H.261

The International Telecommunication Union (ITU) standardized the compression and decompression of digital video in 1990. This standard is called H.261 [26]. Driving force was the need of video telephony and conferencing systems over ISDN. H.261 describes the video encoding and decoding methods for the moving picture component of audio-visual services at the rate of p*64 kbits/s, where p takes values from 1 up to 30, which represents the number of B channels of ISDN, each having a data rate of 64 kbits/s. Therefore, H.261 is also known as "$p \times 64$" [35]. A real-time encoding-decoding with less than 150 ms delay was the focus of the ITU expert group. The standard specifies the decoder, the syntax, and the semantics of the coded bit stream. Whereas JPEG treats its picture independently (intraframe coding), H.261 tries to predict the current picture from the previous one (interframe coding) in order to reduce the picture information to be transmitted. The processing steps of a H.261 video codec are depicted in Figure 3–14. See the standard [26] for more details.

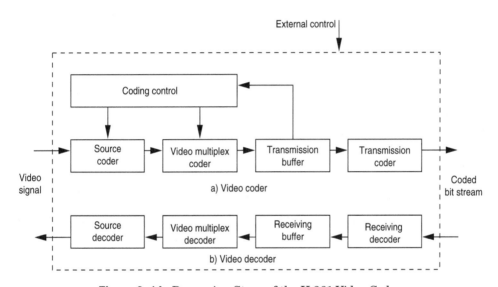

Figure 3–14 Processing Steps of the H.261 Video Codec

Image Preparation in H.261

The image model of H.261 is very restricted compared with that of JPEG. An image in the H.261 model consists of three rectangular matrices with 8 bits per sample: one luminance matrix Y and two chrominance matrices C_b and C_r for the blue and red color components of each pixel, respectively. The chrominance matrices have half the horizontal and half the vertical resolution of the luminance matrix (4:1:1 format). The location of luminance and chrominance samples within an uncompressed frame is depicted in Figure 3–15.

An H.261 encoder processes source images only in CIF (Common Intermediate Format) or QCIF (Quarter CIF); see Table 3–3. All H.261 implementations must be able to encode and decode QCIF; ability to encode and decode CIF is optional. CIF and QCIF frames are divided into a hierarchical block structure consisting of picture, groups of blocks (GOB), macro blocks, and blocks (see also Table 3–4). Each macro block is composed of four 8×8 luminance blocks and two 8×8

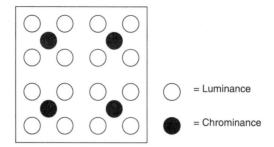

Figure 3–15 Location of Luminance and Chrominance Samples in H.261

chrominance blocks. A GOB is composed of 3×11 macro blocks. A QCIF picture has three GOBs; while its CIF equivalent has four times that number. Picture processing and entropy encoding are performed on macro blocks.

	CIF	**QCIF**
	($width \times height$)	($width \times height$)
Y	352×288	176×144
C_b	176×144	88×72
C_r	176×144	88×72

Table 3–3   Source Image Formats of H.261 (in Samples)

Structural Element	**Description**
Picture	1 video picture (frame)
Group of blocks	33 macro blocks
Macro block	16×16 Y, 8×8 C_b, C_r
Block	8×8 pixels (coding unit for DCT)

Table 3–4   Hierarchical Block Structure of H.261

Picture Processing in H.261

H.261 knows two types of coding macro blocks, namely, intraframe and interframe. In the case of intraframe coding, no advantage is taken of the redundancy between frames. Beyond this, H.261 tries to make use of temporal redundancies by means of motion-compensated prediction. If the motion estimation is successful and the prediction error is below a certain threshold, interframe coding is applied for the macro blocks.

The first frame to be transmitted is always an intraframe coded frame, i.e., all macro blocks are intraframe coded. The image is coded blockwise. That is, the entire picture is divided into nonoverlapping 8×8 pixel blocks on which, first, the forward DCT is applied. Then, the resulting 64 DCT coefficients are quantized and zigzag-reordered. In preparation for the interframe coding, the recently coded frame is decoded again within the encoder, using inverse quantization and inverse DCT. This decoding is done to obtain exactly the same reference frame as the decoder.

For the next frame to be encoded, the last previously coded and stored frame is used for deciding whether to intraframe- or interframe-code each macro block. The algorithm performs a unidirectional motion-compensated prediction as described in section 3.2.6. The algorithm uses only the four luminance blocks of each macro block, specifying a 16×16 pixel area, to find a close match in the previous frame for the macro block currently encoded. The search area for the motion vector is at most $\pm$ 15 pixels in x- and y-direction. The motion compensation unit of H.261 tries to detect motion by checking macro blocks. If it cannot find a close match, it employs exactly the same coding for the macro block as in intraframe coding. Motion vectors are coded differentially with the motion vector of the macro block to the left used as a prediction. Note that according to the standard, the coder need not to be able to determine a motion vector, i.e., a simple H.261 implementation considers only the differences between macro blocks located at the same position in consecutive frames.

Therefore, the motion estimation process can result in one of three possible decisions for the coding of a macro block:

- Intracoding, where the original intensity values are transform coded
- Intercoding without motion compensation, i.e., the motion vector has zero value
- Intercoding with motion compensation

Also optional is a filter between DCT and the entropy coding process, which can be used to improve the image quality by removing high-frequency noise as needed. This is especially useful in the case of low bit rates.

Blocks can be skipped, i.e., are not coded, if the difference between the current and the predicted block is less than a certain threshold. If it is the same for all six blocks of a macro block that is to be intercoded, the whole macro block may be skipped. It is even possible to skip up to three frames between two coded frames, but at least once every 132 transmitted frames, a macro block should be intracoded to alleviate error propagation.

Unlike JPEG and MPEG-1 Video, H.261 does not make use of quantization matrices. It just applies quantization factors to GOBs. Quantization in H.261 is a linear function, i.e., does not distinguish between DCT coefficients with high and low values. The quantization step size depends on the amount of data in the transform buffer, thereby enforcing a constant data rate at the output of the coder. Therefore, the quality of the encoded video data depends on the contents of individual images as well as on the motion within the respective video scene. Quantization of DC coefficients differs from that of AC coefficients.

Entropy Encoding in H.261

H.261 represents a run of zeros and the non-zero value by just a pair of values (run and amplitude). A variable-length code entry exists for the combinations most likely to occur. Otherwise, a fixed-length code (20 bits) is used, consisting of a 6-bit escape code, 6 bits run, and 8 bits amplitude. All Huffman tables used are predefined by the standard; it defines tables for motion vectors, quantized DCT coefficients, etc.

This section has presented an example of interframe coding, namely, motion-compensated prediction. In the next section, dedicated to MPEG-1 Video, this interframe coding is supplemented by motion-compensated interpolation, which is used by MPEG-1 Video beyond motion-compensated prediction.

3.3.4   MPEG-1 Video

MPEG-1 Video belongs to a family of ISO/IEC standards [21]. The *Moving Picture Experts Group* (MPEG) has defined a bit-stream representation for synchronized digital audio and digital noninterlaced video, compressed to fit into a bandwidth of not more than 1.5 Mbps [34].

About 1.1 Mbps are for video, 128 kbps are for audio, and the remainder are for the MPEG system. A major application of MPEG-1 is the storage of audio-visual information on storage media such as CD-ROM and DAT, which have the corresponding data retrieval speed.

MPEG-1 consists of several parts, of which *System*, *Video*, and *Audio* primarily interest us. The MPEG-1 System is responsible for multiplexing and synchronizing one video stream with one or multiple audio streams. Beyond simple playback, the MPEG system streams allow random access, fast-forward, and rewind. The video part is discussed in this section. MPEG-1 Audio is presented in section 3.4.2

In the development of the MPEG-1 standard, H.261 and JPEG were considered. Like JPEG, MPEG-1 is a generic coding standard for many digital video implementations, i.e., independent of a certain application and constructed like a toolkit. An implementation must therefore not realize the complete standard. The standard defines the syntax and the semantics of the video bit stream that must be decodable at the decoder end.

Random access and high compression ratios are features that cannot be optimized both at the same time. Best random access can be achieved with compression methods for single images since each image in the bit stream can be accessed directly. Thus, the compression ratio is relatively low since temporal redundancies between successive frames are not taken into account. A typical example of such a compression method is Motion-JPEG. On the contrary, H.261 allows either no random access at all or very restricted random access, thereby increasing the compression ratio. To offer a compromise between random access and high compression ratio, MPEG-1 Video distinguishes between four different coding types for images [48], described below: I frames, P frames, B frames, and D frames.

I frames (intracoded frames) are coded without any reference to other images. MPEG makes use of JPEG for I frames. They are points for random access in MPEG streams and can be used as a reference for the coding of other images. The compression rate for I frames is the lowest of all defined coding types.

P frames (predictively coded frames) require information from the previous I and/or P frame for encoding and decoding. By exploiting temporal redundancies, the achievable compression ratio is considerably higher than that for I frames. P frames can be accessed only after the referenced I or P frame has been decoded.

B frames (bidirectionally predictively coded frames) require information from the previous and following I and/or P frames for encoding and decoding. The highest compression ratio is attained by using these frames because, in addition to the coding methods of the other coding types, a bidirectional motion-compensated prediction can be used. B frames can never be used as a reference for the coding of other images. An example of the usefulness of B frames is depicted in Figure 3–16. The body in the I frame is hidden by the closed door; therefore, unidirectional motion-compensated prediction is not practicable. But proceeding from the following P frame, at least half the body can be predicted.

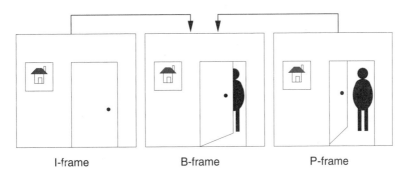

I-frame   B-frame   P-frame

Figure 3–16 Use of B Frames in MPEG Video

D frames (DC coded frame) are encoded intraframe, whereby the AC coefficients are neglected. They can be used for the fast-forward or rewind mode. D frames can never be used with the other picture types, i.e., in order to use fast-forward or rewind mode, both IPB-coded and D-coded video streams must be created and stored.

Reference frames must be transmitted first. Therefore, the transmission order and the display order may differ, as depicted in Figure 3–17. At the beginning, there is always an I frame. The first I frame and the first P frame serve as a reference for the first two B frames. At the same time, the first I frame is also the

reference for the P frame. Therefore, the I frame must be transmitted first, followed next by the P frame and then by the B frames. Thereafter, the second I
frame must be transmitted since it serves as reference for the second pair of B
frames.

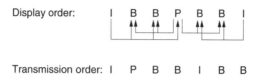

Figure 3–17 Display and Transmission Order in MPEG-1 Video

An important data structure in the data hierarchies of MPEG-1 Video is the
group of pictures (GOP) (see Figure 3–18). A GOP contains a fixed number of consecutive frames and guarantees that the first picture of each GOP is an I frame.
A GOP gives an MPEG encoder information as to which picture should be coded
as an I, P, or B frame and which frames should serve as references. For example,
to access an encoded B frame, at most the GOP and the first picture of the next
GOP, an I frame, must be decoded. Consequently, the GOP can be used for simple
editing operations on MPEG video streams.

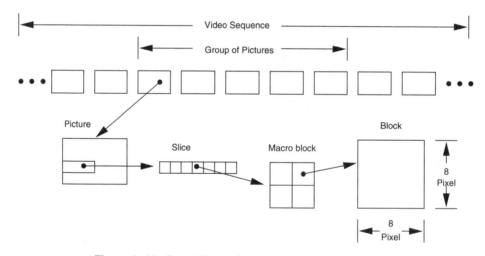

Figure 3–18 Data Hierarchies of MPEG Video Streams

Image Preparation in MPEG-1 Video

MPEG-1 Video uses the same image format as H.261 but allows a greater choice
of image size. The luminance component should not exceed a resolution of
768×576. See also Table 3–5.

Picture Processing in MPEG-1 Video

An MPEG-1 video encoder processes many different image formats. Two *Standard Interchange Formats* (SIF) are defined, namely, PAL (352×288) and NTSC (352×240), derived from the two, well-known television standards. General restrictions are set by the *Constrained Parameter Set* (CPS); see Table 3–5.

Parameter	Restrictions
Horizontal resolution	≤ 768 pixel
Vertical resolution	≤ 576 lines
Macro blocks/s	≤ 25 macro blocks/s
Frames/s	≤ 30 Hz
Motion vector range	$\leq (-64 / + 63{,}5)$ pixel
Input puffer size	≤ 327.680 bit
Bit rate	$\leq 1*856$ Mbps

Table 3–5 The MPEG-1 Constrained Parameter Set

The first frame in each GOP is always an I frame, which is encoded blockwise like an intraframe-coded image in H.261, i.e., with DCT, quantization, and entropy coding. Unlike the case in H.261, the reference frame for a P frame is not necessarily the last previously coded frame. The reference frame or frames are decoded again within the encoder, using inverse quantization and inverse DCT. As with H.261, just the four luminance blocks of a macro block are involved in the motion estimation process.

The use of the motion estimation process depends on the structure of the video sequence, which is defined by the application-given GOP. The motion estimation step is activated when B or P frames appear in the GOP. This is the most common case since these coding types achieve the best compression ratios. The MPEG-1 standard does not specify any limitations for coordinate values of the motion vector. If a P frame must be encoded, the motion estimation unit uses the last I or P frame as the reference frame. If a good match between a macro block in the current frame and a macro block in the reference frame has been found, then a motion vector and the resulting prediction error are encoded for the macro block in the current frame. Otherwise, the macro block is simply intracoded.

The decision process for B frames is more complex. Four possibilities must be taken into account: forward prediction, backward prediction, interpolation, or, if none of these is applicable, intracoding of the macro block (see also section 3.2.6). If interpolation is applied, two reference frames must be available: the closest past and future I or P frame, yielding to two motion vectors and one prediction error block. The reference frames for P and B frames must be transmitted first.

The MPEG quantizer uses different quantization matrices for intraframe-coded macro blocks and for any other macro block because interframe-coded DCT coefficients have statistical properties different from those of intraframe-coded DCT coefficients. The coding control can adapt the output bit rate of the encoder by means of a Q factor. Bit-rate requirements can thereby be satisfied, and the picture quality improved as much as possible. In the event of a transmission error, the remaining macro blocks of a slice are replaced by skipped macro blocks until a new slice starts. Unlike the case in H.261, picture skipping is not allowed.

Entropy Encoding in MPEG-1 Video

Entropy encoding is mainly the same as in H.261. It is a two-stage process in which run-length codes of variable and fixed length are first applied, followed by Huffman coding with predefined tables. The variable-length code associated with DCT coefficients is a superset of the one used in H.261 to reduce costs when implementing both standards in a single processor [34].

A performance test of JPEG, H.261, and MPEG-1 Video is considered by Milde [37]. A brief summary of Milde's results, focused on H.261 and MPEG-1 Video, can be found in the work of von Roden [45].

3.3.5 MPEG-2 Video

In this section, we discuss major extensions of MPEG-1 Video embodied in MPEG-2 Video. See also [23] and [49] for more information.

MPEG-2 is, in conformity with MPEG-1, a method for the compression of digital audio-visual signals. The video part of MPEG-2 permits data rates up to 100 Mbits/s and also supports interlaced video formats and a number of advanced features, including those supporting HDTV. MPEG-2 Video can be used for the digital transmission of video over satellite, cable, and other broadcast channels. It builds upon the completed MPEG-1 standard and was cooperatively developed by ISO/IEC (IS 13818-2 [23]) and ITU (H.262 [27]).

As a generic standard, MPEG-2 Video was defined in terms of extensible profiles, each of which supports the features needed by an important class of applications. The MPEG-2 *main profile* was defined to support digital video transmission in the range of about 2 to 80 Mbps. Parameters of the main profile and the *high profile* are suitable for supporting HDTV formats. In addition, a *hierarchical/scalable profile* was defined to support applications such as compatible terrestrial TV/HDTV, packet-network video systems, and other applications for which multilevel coding is required. In addition, backward compatibility with existing standards (MPEG-1 and H.261) has been defined.

All profiles are arranged in a 5×4 matrix, as shown in Table 3–6 (adapted from Teichner [50]). The horizontal axis denotes profiles with an increasing number of operations to be supported. The vertical axis indicates levels with increased parameters. The upper bound for the sampling density (pixels/line $\times$

lines/frames × frames/s) for each level is also stated. For example, the main profile in the high level has an upper bound of 1920 pixels/line, 1152 lines/frames, and 60 frames/s with a data rate less than or equal to 80 Mbps.

	Simple Profile (no B frames, not scalable)	Main Profile (B frames, not scalable)	SNR Scalable Profile (B frames, SNR scalable)	Spatially Scalable Profile (B frames, spatial or SNR scalable)	High Profile (B frames, spatial or SNR scalable)
High level (1920×1152 ×60)		≤80 Mbps			≤100 Mbps
High-1440 level (1440×1152 ×60)		≤60 Mbps		≤60 Mbps	≤80 Mbps
Main level (720×576×30)	≤15 Mbps	≤15 Mbps	≤15 Mbps		≤20 Mbps
Low level (352×288×30)		≤4 Mbps	≤4 Mbps		

Table 3–6 MPEG-2 Video Profiles and Levels with the Most-Important Characteristics (Cells in the table without entries are not defined)

To encode interlaced video signals effectively, MPEG-2 Video extends the motion-compensated prediction of MPEG-1 Video and allows other sequences of DCT coefficients. In addition, algorithms for scalable and hierarchical coding are defined, thereby making possible the use of the same data stream for both normal TV and HDTV systems. This feature is also useful in a multicast scenario: in each subtree in a multicast distribution, only those parts of the signal that can be used by the receivers in that subtree need be transmitted.

Initially, MPEG-3 was intended to support HDTV applications. During development, however, MPEG-2 Video proved adequate when scaled up to meet HDTV requirements [48]. Consequently, MPEG-3 was dropped. The next section is dedicated to MPEG-4 Video, which supports low-bit-rate applications.

3.3.6 MPEG-4 Video

Recently, the ISO expert group developing the MPEG-4 standard has decided to stop the development of a new video coding method for low bit rates. Instead, they focus on providing enhanced functionality based on existing compression methods. An example of such enhanced functionality is the coding of audio-visual objects, i.e., coding of individual objects with any shape in a video scene. Consider, for example, a video clip showing a tennis match. Instead of each image in the clip coded as a whole, the stationary background and the tennis player in the foreground can be coded independently with different methods or parameter sets. On the audio side, audio objects are identified and coded depending on their contents. Examples of different audio contents are music, speech, and other sensible sounds.

One of the existing video coding methods under study for MPEG-4 is H.263, described in the next section. An enhanced version of H.263 has been developed to increase visual quality at low bit rates by removing blocking artifacts without reducing the sharpness [51].

3.3.7   H.263

The *ITU-T Recommendation H.263* defines a codec for the compression of the moving picture component of audio-visual services at low bit rates [28]. A typical application is the transmission of video over a V.34 modem connection, using 20 kbps for video and 6.5 Kbits for audio. The recommendation is based upon H.261 and extends this standard in several areas: First, instead of two image formats, five are now supported. Second, the motion-compensated prediction has been refined; and, third, the standard now supports B frames. In contrast to MPEG Video, B frames in H.263 can have only P frames as a reference. The Recommendation H.263 improves H.261 and takes into account the experience of the MPEG-Video standard.

In these last sections, up-to-date methods of video compression have been presented. In the following two sections, we outline fractal and wavelet image compression as examples of next-generation coding methods. Both will attract higher interest in the future.

3.3.8   Wavelet Image Compression

The Fourier transform reveals information only about the frequency domain behavior of the signal. It is not well suited to also obtain information about the time behavior of the signal [16]. In the last few years, wavelet transform has become a cutting-edge technology both in compression and signal analysis research. One important reason behind this development is that discrete wavelet transform can decorrelate signals, e.g., in the form of image data, into low- and high-frequency events.

Though DWT is also used in audio compression, most of the recent work on DWT focused on image signal analysis and image compression. See [55] for an introduction to wavelet audio compression.

But what is a *wavelet*? A wavelet is defined as a set of basis functions, derived from the same prototype function. The prototype function is also known as the mother wavelet. Two examples of mother wavelets are shown in Figure 3–19. Each of the basis functions has a finite support of a different width and can be scaled and transformed to meet the requirements of an application. The different support allows different trade-offs of time and frequency resolution. To resolve low-frequency details accurately, a wide basis function can be used to examine a large region of the signal, whereas to resolve time details accurately, a short basis function should be used to examine a small region of the signal.

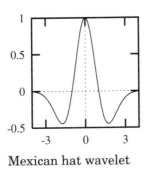

Mexican hat wavelet

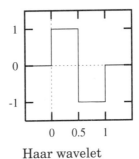

Haar wavelet

Figure 3–19 Two Mother Wavelets

In contrast to DCT-based coding methods, which remove most of the high-frequency information from the data stream, wavelet transform coders process the high- and low-frequency parts of the signal independently. Most of the energy in the high-frequency portions of images is attributable to edges, both isolated edges and clusters of edges found in textured regions. Edges are important for recognition of the content of an image, but not all edges are equally important for the human visual system. Depending on the required compression ratio, more and more information about edges can be removed, thereby degrading image quality.

In addition, using wavelet transform, an image is transformed as a whole and not subdivided into pixel blocks, as with DCT-based coding methods. Therefore, no blocking artifacts occur. Instead, even at high compression ratios, wavelet coders degrade gracefully [16]. The left image in Figure 3–20, compressed to 1:16 of its original size, shows very good image quality with no visual artifacts. At a compression level of 1:256, the right image in Figure 3–20 shows a graceful degraded image quality instead of blocking artifacts, as with DCT-based coding methods (compare with the right image in Figure 3–7).

As mentioned above, wavelet transform is applied to complete images. Figure 3–21 depicts a block diagram of a forward wavelet transform. The image is first filtered along the x-dimension, resulting in a low-pass and a high-pass image. Since the bandwidth of both the low-pass and the high-pass image is now half that of the original image, both filtered images can be downsampled by a factor

Lena image (1:16)   Lena image (1:256)

Figure 3–20 Examples of Artifacts When Using DWT-based Coding Methods

of 2 without loss of information. Then, both filtered images are filtered and down-sampled along the y-dimension, resulting in four subimages. One of these subimages represents the average signal, and the three remaining images represent the horizontal image features, the vertical features, and the diagonal features (see also Figure 3–5).

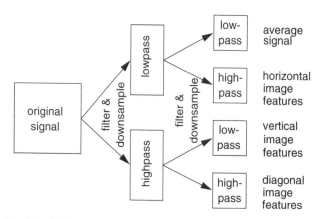

Figure 3–21 Block Diagram of Two-Dimensional Forward Wavelet Transform

The average signal can be recursively transformed for a better compression ratio. After the forward wavelet transform, compression has not yet been accomplished. Instead, each application of the forward wavelet transform increases the number of coefficients, i.e., the storage requirement increases. Compression is achieved, as with DCT, by quantizing and entropy encoding the wavelet coefficients. By inverting the coding operations, data can be reconstructed. See [8][16][36] for more information on wavelet transform.

Both DCT and DWT are symmetrical coding methods. Because of the better localization of frequencies of DWT, better compression ratios at the same visual quality as that with DCT can be achieved. Efficient implementations have been developed for both DCT and DWT, with a small advantage for DCT because of experience gained in that field; but this will change in the future. One of the most important advantages of DWT over DCT is the inherent scalability of DWT by a factor of 2. In DCT-based coding methods, this scalability is achieved by adding functionality, as within the hierarchical JPEG mode, resulting in a lower compression factor caused by the additional overhead.

The image compression method described in this section can be extended to compress sequences of digital images by exploiting temporal redundancies in one of two ways [16]: First, standard video compression techniques like hierarchical motion compensation and three-dimensional subband coding can be implemented using wavelets, thereby extending the two-dimensional DWT by the time dimension. Second, instead of performing the complete inverse DWT for each frame in a slowly varying image sequence, one needs to compute only the inverse DWT for those pixels that have changed by a meaningful amount between adjacent frames in the sequence. Experiments have shown that higher compression rates can be achieved by the first approach than by the second. However, the memory and processor requirements of the first approach are significantly higher than those of the second, which can be implemented very efficiently.

3.3.9 Fractal Image Compression

Totally different from transform coding is the approach of fractal image coding. Fractal block coders exploit image redundancy through self-transformability on a blockwise basis. They store images as contraction maps of which the images are approximate fixed points [9]. This transformation leads to a representation of an image as a fractal, an object with details in all scales, i.e., fractal image coding takes advantage of image redundancy in scale. In principle, the techniques described in this section can also be applied to audio signals, but not much research has been performed in that field. Therefore, we concentrate on fractal image coding as described by Fisher [13] and Barnsley and Hurd [3].

Transform coders based on discrete cosine transform or wavelet transform are designed to take advantage of very simple structures in images, namely, on the fact that values of pixels that are close together are correlated. They eliminate high-frequency and low-frequency components of a signal to such an extent that successive coding steps can compress the data. In contrast to that, fractal image coding is motivated by the observation that important image features, such as straight edges and constant regions, are invariant under rescaling.

This local scale invariance is exploited by fractal block coders through use of coarse-scale image features to quantize fine-scale features. Fractal image compression is related to vector quantization, but it uses a self-referential vector codebook, drawn from the image itself, rather than a fixed codebook. Images are not stored as a set of quantized transform coefficients but, instead, as fixed points of maps on the plane.

The codebook is constructed by the coder from locally averaged and sub-sampled isometrics of larger blocks from the image. This codebook is effective for coding constant regions and straight edges because of the scale invariance of these features. The coder determines a self-referencing contraction map of a plane of which the image to be coded is an approximate fixed point. Images are stored by saving the parameters of this map and decoded by iteratively applying this map to find its fixed point. By effectively storing only the parameters of the contraction map, the image is compressed.

In general, transformations can skew, stretch, rotate, scale, and translate an image block. In addition, contrast and brightness of the transformation can be changed. The transformation process of mapping coarse-scale image features to fine-scale features is depicted in Figure 3–22 (derived from Davis [10]). In this example, the original block in the image on the left side is first averaged and sub-sampled, and then rotated. Finally, the contrast is modified, and the addition of an offset adjusts the transformed image block.

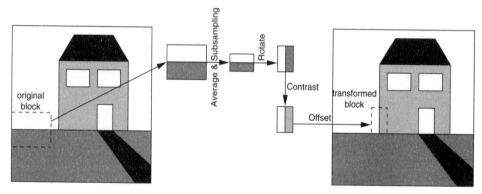

Figure 3–22   Transformation of an Image Block

It is clear that fractal block coders are effective for images composed of iso-lated straight lines and constant regions, since these features are self-similar. In general, images must be composed of features at a fine scale that are also present at a coarser scale up to rigid motion and transforms of intensities. This is the "self-transformability" described by Jacquin [29]. But this assumption also holds when more complex features are present because complex image features such as textures tend to possess characteristics that can be exploited by fractal coders [10]. Pixel values in natural images are far from independent, especially values for pixels that are close together.

Fractal image coding is a lossy coding method since natural images are not exactly self-similar. The decoded image is only an approximation of the original, but if the transformations are carefully chosen, the difference between the approximation and the original image is hard to detect at low compression ratios. Figure 3–23 shows as examples on the left side the Lena image compressed to 1:16 of its original size and on the right side compressed to 1:256 of its original size (compare with the original Lena image on the left side in Figure 3–7).

Lena image (1:16)   Lena image (1:256)

Figure 3–23   Examples of Artifacts When Using a Fractal Image Coder

One of the main problems of fractal image compression is the determination of the concrete contraction maps, i.e., the determination of the codebook, that can be used to effectively code an image. For that purpose, *iterated function systems* (IFS [2]) are used. An IFS is a finite set of functions used by a fractal coder to look systematically for self-similarity within an input image by scaling, rotating, skewing, or translating image blocks.

Fractal image compression is increasing in popularity. It is used in some commercial applications, e.g., the Microsoft Encarta CD-ROM and the Fractal Imager of Iterated Systems, and in many public domain implementations. A great variety of information and resources on fractal image coding and related topics can be found on a Web page edited by Yuval Fisher (URL= http: // inls3.ucsd.edu:80/Research/Fisher/Fractals/).

3.4 Audio Compression

In the previous sections, we concentrated on compression methods for digital video. In the following sections, we focus on compression methods for digital audio. Only two dimensions need be taken into account by coding methods for digital audio streams, namely, amplitude and time. At first sight, these two dimensions reduce complexity compared to the three dimensions: width, height, and time of digital video. But the human auditory system is much more sensitive to quality degradation than is the human visual system, i.e, the amount of redundancy that can be removed for compression is relatively small. Therefore, even the most advanced coding methods for digital audio have compression ratios far below those for digital video. Moreover, separation of relevant from irrelevant information often requires much computation, as we will show.

Audio compression methods differ in the trade-offs between encoder and decoder complexity, the quality of the compressed audio, and the amount of data compression [56]. The following sections present examples which cover the full range from low-complexity, low-compression, and medium audio quality algorithms up to high-complexity, high-compression, and high audio quality algorithms. These techniques apply to general audio signals, i.e., are not optimized for specific audio signals like speech. See [43] for more information on speech coding.

3.4.1   Variants of Pulse Code Modulation

The data rates associated with PCM-coded audio streams are substantial. A basic audio compression technique is employed in digital telephony. This method is based on a transformation which is logarithmic in nature; it maps 13 or 14 bits of linearly quantized PCM values to 8-bit codes. The mapping from 13 to 8 bits is known as A-*law* transformation, and the mapping from 14 to 8 bits is standardized as μ-*law* transformation. Unlike linear quantization, the logarithmic step spacing represents low-amplitude audio samples with greater accuracy than it does higher amplitude values. Thus, the signal-to-noise ratio of the transformed output is more uniform over the range of amplitudes of the input signal. The A-law transformation is in common use in Europe for ISDN 8-kHz-sampled, voice-grade, digital telephony service, and the μ-law transformation is used in North America and Japan. The ITU recommendation G.711 specifies both A-law and μ-law transformations [25].

Adaptive Differential Pulse Code Modulation (ADPCM) has been developed to overcome the disadvantages of DPCM, as described in section 3.2.5 ADPCM is a lossy audio-coding method that codes differences between PCM-coded audio signals, using only a small number of bits (e.g., 4 bits) but can adapt to the characteristics of the signal by changing the step size of the quantizer, the predictor, or both. Given such a small number of bits, either the high- or the low-frequency portion of a signal can be coded exactly. Therefore, an ADPCM coder always operates in one of two modes: high- or low-frequency mode. In either of these two modes, the other part of the frequency portion of the signal is almost ignored. These techniques reduce the data rate of high-quality audio from 1.4 Mbps to 32 kbps. ADPCM is standardized as CCITT recommendation G.721 [6].

3.4.2   MPEG-1 Audio

The data rates attained by the methods described above are still too high to permit storage and transmission of high-fidelity audio. Therefore, the Motion Pictures Expert group has developed two international standards for the compression of audio signals: MPEG-1 Audio [22] and MPEG-2 Audio [24]. In our discussion, we will concentrate on MPEG-1 Audio. The extension of MPEG-2 Audio beyond MPEG-1 Audio is outlined at the end of this section.

Like μ-law and ADPCM, MPEG-1 Audio compression is lossy. However, the MPEG algorithm can achieve transparent, perceptually lossless compression, as has been proved by extensive subjective listening tests during the development of the standard [39]. The MPEG algorithm exploits perceptual limitations of the human auditory system, namely, the hearing threshold and auditory masking, to determine which part of an audio signal is acoustically irrelevant and can be removed for compression.

The human ear has a limited frequency selectivity that varies in acuity from less than 100 Hz for the lowest audible frequencies to more than 4 kHz for the highest. This *hearing threshold* of the human ear is roughly depicted in Figure 3–24. Thus, the audible spectrum can be partitioned into critical bands that reflect the resolving power of the ear as a function of frequency.

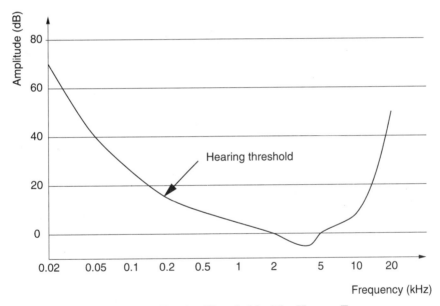

Figure 3–24 Hearing Threshold of the Human Ear

Auditory masking is a perceptual weakness of the ear that occurs whenever the presence of a strong audio signal makes a spectral neighborhood of weaker audio signals imperceptible. The threshold for noise masking at any given frequency is solely dependent on the signal activity within a critical band of that frequency. This audio noise-masking property is illustrated in Figure 3–25 (derived from [39]).

Thus, it is the task of an MPEG-Audio coder to identify and remove perceptually irrelevant parts of the audio signal. An MPEG-Audio coder works by divid-

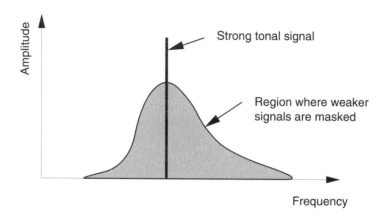

Figure 3–25 Audio Noise-Masking Property of the Human Auditory System

ing the audio signal into frequency subbands that approximate critical subbands and then quantizing each subband according to the audibility of quantization noise within that band with as few levels as possible.

The MPEG-1-Audio standard defines three distinct layers of compression; each implementation of a higher layer must be able to decode the MPEG audio signals of lower layers, i.e., a layer III decoder can also process bit streams generated by a layer I or a layer II encoder. Each successive layer improves the compression performance but at the cost of greater encoder and decoder complexity. Figure 3–26 and Figure 3–27 show block diagrams of the MPEG audio encoder and decoder, respectively (derived from [4]). The input audio stream passes simultaneously through a filter bank and through a *psychoacoustic model*. The filter bank divides the input into multiple subbands, and the psychoacoustic model determines the signal-to-mask ratio of each subband. These ratios are used by the bit or noise allocation block; it determines the number of bits needed to make the quantization noise at each subband inaudible. The last step formats the quantized samples into a decodable bit stream. At the decoder side, the formatting is reversed, then the quantized subband values are reconstructed and finally transformed into a time-domain signal.

Each layer defines fixed bit rates for the coded audio stream; only layer III allows variable bit rates. The minimal value is always 32 kbps. Layer I is the simplest. Its algorithm uses the basic filter bank found in all layers. This filter bank divides the audio signal into 32 constant-width frequency bands by means of a fast Fourier transform. Layer I uses a relatively simple psychoacoustic model because its main target is easy implementation. Therefore, this layer best suits bit rates above 128 kbps. For example, Philips' Digital Compact Cassette (DCC) uses Layer I compression at 192 kbps per channel.

The layer II algorithm is a straightforward enhancement of layer I. It has a target bit rate of 128 kbps per channel. Its psychoacoustic model is somewhat more complex than that of layer I. In addition, it uses a more efficient code for the representation of the bit allocation, the scale factor values, and the quantized val-

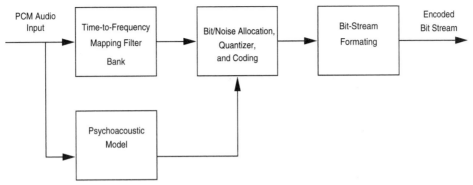

Figure 3–26 MPEG-1 Audio Encoder

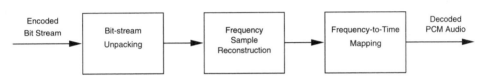

Figure 3–27 MPEG-1 Audio Decoder

ues; therefore, more code bits are available to represent the quantized subband values. Possible applications for this layer are the storage of audio sequences on CD-ROM and the audio track of the Video-CD [39].

Layer III is the most complex but offers the best audio quality, particularly for bit rates around 64 kbps per channel. Therefore, this layer can be used for the transmission of high-fidelity audio over ISDN. In addition to division of the audio signal into 32 subbands described above, this layer compensates for some filter-bank deficiencies by processing the filter outputs with a modified discrete cosine transform, resulting in 576 so-called lines. These lines are used for further processing. The psychoacoustic model of this layer is more complex than that of both layer I and layer II to accommodate the high requirements of this layer, one of whose unique features is the additional usage of Huffman codes for entropy coding.

The MPEG-1-Audio standard defines a bit stream that can support one or two audio channels: a single channel, two independent channels, or one stereo signal. The two channels of a stereo signal can be processed either independently or as *joined stereo* to exploit stereo redundancy.

The MPEG-2-Audio standard extends the functionality of its predecessor by multichannel coding with up to five channels (left, right, center, and two surround channels), plus an additional low-frequency enhancement channel, and/or up to seven commentary/multilingual channels. It also extends stereo and mono coding of the MPEG-1-Audio standard by further sampling rates. The MPEG-2-Audio standard provides backward compatibility with the MPEG-1-Audio standard, i.e., an MPEG-2-Audio decoder can process any MPEG-1-Audio bit stream.

In addition, an MPEG-1-Audio decoder can read and process the stereo information of an MPEG-2-Audio stream.

3.4.3   Perceptual Audio Coder

MPEG-1 Audio exploits masking in the frequency domain (also called *simultaneous masking*) to separate relevant from irrelevant information. Recently developed methods also exploit masking in the time domain (also called *non-simultaneous masking*) to increase even further the compression rate for digital audio without degrading quality. These methods analyze whether a strong signal in a given time block can mask an adequately lower distortion in a previous or subsequent block (*backward or forward masking*).

An example of such a method is PAC (*Perceptual Audio Coder*), developed by the AT&T Bell Laboratories [31]. At the ISO-MPEG-2 audio test, PAC demonstrated the best decoded audio signal quality of any algorithm at 320 kbps for 5-channel audio. However, PAC is not backward compatible, i.e., it cannot decode MPEG-1 audio streams. PAC can provide audio compression at a stereo coding rate of 128 kbps to 160 kbps.

In the last few years, considerable improvements in the compression of digital audio and video data have been attained, which were not foreseeable in the past. The end of this development has by no means been reached.

3.5 More Information about Compression Methods

The Internet in general and the World Wide Web in particular are the ideal places to find up-to-date information about development and application of compression techniques. Unfortunately, not all important documents can be accessed without fees because standardization organizations like ISO and ITU demand fees for some of their documents.

Nevertheless, by using one of the search engines listed in Table 3–7, one can find status reports, audio and video examples, source code, and much other information on compression methods. But the Web has grown to such an extent that looking for a keyword like "MPEG" or "JPEG" would result in thousands of hits—impossible to check each in reasonable time.

Search Engine	URL
Alta Vista	http://altavista.digital.com/
Lycos	http://lycos.cs.cmu.edu/
Yahoo!	http://www.yahoo.com/

Table 3–7   Search Engines on the Web

Therefore, we provide in Table 3–8 an alphabetically sorted list of URLs, referring to excellent collections of URLs, research groups, and research projects. This list can be used both as a starting point to surf the Web on the hunt for information about compression methods as well as to discover quickly some of the treasures of the World Wide Web.

Description	URL
Emphasis project (MPEG-4)	http://www.fzi.de/esm/projects/emphasis/welcome.html
Fractal image compression	http://inls3.ucsd.edu:80/Research/Fisher/Fractals/
ISO online	http://www.iso.ch/welcome.html
ITU	http://www.itu.ch/
MPEG pointers and resources	http://www.bok.net/~tristan/MPEG/MPEG-content.html
Wavelets	http:// www.mat.sbg.ac.at/~uhl/wav.html

Table 3–8   Hints on Compression Methods on the Web

References

[1]   Apple Computer, Inc., *Inside Macintosh: QuickTime*. Reading, MA: Addison-Wesley, 1993.

[2]   M. F. Barnsley and S. Demko, "Iterated Function Systems and the Global Construction of Fractals," *Proc. Royal Society of London*, A399:243-275, 1985.

[3]   M. F. Barnsley and L. P. Hurd, *Fractal Image Compression*, Wellesley, MA: AK Peters, Ltd., 1993.

[4]   K. Brandenburg and G. Stoll, "The ISO/MPEG-Audio Codec: A Generic Standard for Coding of High Quality Digital Audio," In *Convention of the Audio Engineering Society*, 1992.

[5]   G. Campbell, T. A. DeFanti, J. Frederiksen, S. A. Joyce, L. A. Leske, J. Lindberg, and D. J. Sandin, "Two Bit/Pixel Full Color Encoding," *Computer Graphics*, 20(4):215-223, 1986.

[6]   CCITT. *32 kbit/s Adaptive Differential Pulse Code Modulation (ADPCM) of Voice Frequencies*. Recommendation CCITT G.721, 1988.

[7]   E. Chan, A. A. Rodriguez, G. Rakeshkumar, and S. Panchanathan, "Experiments on Block-matching Techniques for Video Coding," *Multimedia Systems*, 2(5):228-241, Dec. 1994.

[8]   C. K. Chui, "An Introduction to Wavelets," Vol. I of *Wavelets Analysis and its Applications*. San Diego, CA: Academic Press, Inc., 1992.

[9]   G. Davis, "Adaptive Self-quantization of Wavelet Subtrees: A Wavelet-based Theory of Fractal Image Compression," In *SPIE Conference on Mathematical Imaging: Wavelet Applications in Signal and Image Processing*, San Diego, July 1995. URL: http: //www.cs. dartmouth.edu /~gdavis.

[10] G. Davis, A Wavelet-Based Analysis of Fractal Image Compression. URL: http: //www.cs. dartmouth.ed /~gdavis, 1996. (Submitted for publication IEEE TOIP.)

[11] E. J. Delp and O. R. Mitchell, "Image Compression Using Block Truncation Coding," *IEEE Transactions on Communications*, COM-27(9):1335-1342, Sept. 1979.

[12] S. Fischer, R. Lienhart, and W. Effelsberg, "Automatic Recognition of Film Genres," In *Proceedings of ACM Multimedia 95*:295-304 (San Francisco, CA, USA, November 5-9, 1995).

[13] Y. Fisher, editor, *Fractal Image Compression: Theory and Application*. New York: Springer-Verlag, 1995.

[14] J. L. Green, "The Evolution of DVI System Software," *Communications of the ACM*, 35(1):53-67, Jan. 1992.

[15] W. Heise and P. Quattrocchi. *Informations- und Codierungstheorie*. Springer-Verlag, 2nd ed., 1989.

[16] M. L. Hilton, B. D. Jawerth, and A. Sengupta, "Compressing Still and Moving Images With Wavelets. *Multimedia Systems*, 2(5):218-227, Dec. 1994.

[17] D. S. Hirschberg and D. A. Lelewer, "Data Compression," *ACM Computing Surveys*, 19(3):261-296, Sept. 1987.

[18] D. A. Huffman, "A Method for the Construction of Minimum Redundancy Codes," *Proceedings of the Institute of Radio Engineers (IRE)*, 40:1098-1101, Sept. 1952.

[19] A. C. Hung and T. H.-Y. Meng, "A Comparison of Fast Inverse Discrete Cosine Transform Algorithms," *Multimedia Systems*, 2(5):204-217, Dec. 1994.

[20] ISO/IEC. *Information Technology — Digital Compression and Coding of Continuous-Tone Still Images (JPEG)*. International Standard ISO/IEC IS 10918-1, 1992.

[21] ISO/IEC. *Information Technology — Coding of Moving Pictures and Associated Audio for Digital Storage Media up to about 1.5 MBit/s (MPEG) — Part 2: Video*. International Standard ISO/IEC IS 11172-2, 1993.

[22] ISO/IEC. *Information Technology — Coding of Moving Pictures and Associated Audio for Digital Storage Media up to about 1.5 MBit/s (MPEG) - Part 3: Audio*. International Standard ISO/IEC IS 11172-3, 1993.

[23] ISO/IEC. *Information Technology — Generic Coding of Moving Pictures and Associated Audio Information — Part 2: Video*. International Standard ISO/IEC IS 13818-2, 1996.

[24] ISO/IEC. *Information Technology — Generic Coding of Moving Pictures and Associated Audio Information — Part 3: Audio*. International Standard ISO/IEC IS 13818-3, 1995.

[25] ITU-T. *Pulse Code Modulation (PCM) of Voice Frequencies*. Recommendation G.711, 1988.

[26] ITU-T. *Video Codec for Audiovisual Services at p*64 kbit/s*. Recommendation H.261, 1993.

[27] ITU-T. *Generic Coding of Moving Pictures and Associated Audio*. Recommendation H.262, 1994.

[28] ITU-T. *Video Coding for Low Bitrate Communication*. Draft ITU-T Recommendation H.263, 1995.

[29] A. Jacquin, "Image Coding Based on Fractal Theory of Iterated Contractive Image Transformations," *IEEE Transactions on Image Processing*, 1(1):18-30, Jan. 1992.

[30] N. Jayant, J. D. Johnston, and R. J. Safranek, "Signal Compression Based on Models of Human Perception," *Proceedings of the IEEE*, 81(10), Oct. 1993.

[31] N. S. Jayant, E. Y. Chen, J. D. Johnston, S. R. Quackenbush, S. M. Dorward, K. Thompson, R. L. Cupo, J.-D. Wang, C.-E. W. Sundberg, and N. Seshadri, "The AT&T In-Band Adjacent Channel System for Digital Audio Broadcasting," In *Proceedings of International DAB Symposium*, Toronto, Canada, Mar. 1994.

[32] N. S. Jayant and P. Noll. *Digital Coding of Waveforms*. Englewood Cliffs, New Jersey: Prentice Hall, 1984.

[33] B. Lamparter and W. Effelsberg, "eXtended Color Cell Compression – A Runtime-efficient Compression Scheme for Software Video." In *Proc. Multimedia: Advanced Teleservices and High-Speed Communication Architectures,* edited by R. Steinmetz. Berlin: Springer LNCS 868, 1994:181-190.

[34]   D. Le Gall, "MPEG: A Video Compression Standard for Multimedia Applications," *Communications of the ACM*, 34(4):46-58, April 1991.

[35]   M. Liou, "Overview of the p*64 kbit/s Video Coding Standard," *Communications of the ACM*, 34(4):59-63, Apr. 1991.

[36]   S. G. Mallat, "A Theory For Multiresolution Signal Decomposition: The Wavelet Representation," *IEEE Transactions Of Pattern Analysis And Machine Intelligence,* 11(7):674-693, 1989.

[37]   T. Milde. *Video Compression Methods in Comparison: JPEG, MPEG, H.261, XCCC, Wavelets, Fractals* (in German). dpunkt - Verlag für digitale Technologie GmbH, Heidelberg, 1995.

[38]   B. K. Neidecker-Lutz and R. Ulichney, "Software Motion Pictures," *Digital Technical Journal,* 5(2), 1993.

[39]   D. Pan, "A Tutorial on MPEG/Audio Compression," *IEEE MultiMedia*, 2(2):60-74, Summer 1995.

[40]   W. B. Pennebaker and J. L. Mitchell, *JPEG Still Image Data Compression Standard*. New York: Van Nostrand Reinhold, 1993.

[41]   H. A. Peterson, H. Peng, J. H. Morgan, and W. B. Pennebaker, "Quantization of color image components in the DCT domain," In *Proc. SPIE 1453*, pages 210-222, 1991.

[42]   S. Pfeiffer, S. Fischer, and W. Effelsberg, *Automatic Audio Content Analysis*. Technical Report TR-96-008, Praktische Informatik IV, Universität Mannheim, Apr. 1996. URL: ftp: //pi4.informatik.uni-mannheim.de /pub/ techreports/tr-96-008.ps.gz.

[43]   L. Rabiner and B.-H. Juang, *Fundamentals of Speech Recognition*. Englewood Cliffs, NJ: Prentice Hall, 1993.

[44]   K. R. Rao and P. Yip, *Discrete Cosine Transform: Algorithms, Advantages, and Applications*. San Diego: Academic Press, Inc., 1990.

[45]   T. v. Roden, "H.261 and MPEG1 - A Comparison," In *Int. Phoenix Conference on Computers and Communication (IPCCC96)*. IEEE, Mar. 1996.

[46]   A. A. Rodriguez and M. Ken, "Evaluating Video Codecs," *IEEE MultiMedia*, 1(3):25-33, Fall 1994.

[47]   R. Steinmetz, "Data Compression In Multimedia Computing – Principles and Techniques," *Multimedia Systems*, 1(4):166-172, 1994.

[48]   R. Steinmetz, "Data Compression In Multimedia Computing – Standards And Systems," *Multimedia Systems*, 1(5):187-204, 1994.

[49]   R. Steinmetz and K. Nahrstedt, *Multimedia: Computing, Communications and Applications.* Upper Saddle River, NJ: Prentice Hall, 1995.

[50]   D. Teichner, "Der MPEG-2-Standard: Generische Codierung für Bewegt-bilder und zugehöhriger Audio-Information - MPEG-1 und MPEG-2: Universelle Werkzeuge für Digitale Video- und Audio-Applikationen (Teil 1)." *Fernseh- und Kino-Technik,* 48(4):155-163, 1994.

[51]   Telenor R&D. *Enhanced H.263.* URL: http: //www.fou.telenor.no /brukere/ DVC/mpeg4 /H.263+.html, 1995.

[52]   G. K. Wallace, "The JPEG Stillpicture Compression Standard," *Communications of the ACM,* 34(4):30-44, Apr. 1991.

[53]   J. Watkinson, *The Art of Digital Audio.* Oxford, London & Boston: Focal Press, 1988.

[54]   J. Watkinson *The Art of Digital Video,* 2nd ed., Oxford, London & Boston: Focal Press, 1995.

[55]   M. V. Wickerhauser, "Acoustic Signal Compression with Wavelet Packets," In *Wavelets: A Tutorial in Theory and Applications,* Vol. II of *Wavelet Analysis and its Applications,* edited by C. K. Chui. San Diego: Academic Press, Inc., 1992:679-701.

[56]   D. Yen Pan, "Digital Audio Compression," *Digital Technical Journal,* 5(2), 1993.

[57]   J. Ziv and A. Lempel, "A Universal Algorithm for Sequential Data Compression," *IEEE Transactions on Information Theory,* 23(3):405-412, July 1978.

Subnetwork Technology[*]

Heinrich J. Stüttgen

In this chapter, we will analyze how well different networking technologies satisfy the requirements of distributed multimedia applications [17]. To do that, we will first put some numbers to the general performance parameters developed in Chapter 2. Then, we will use these values as a yardstick against the different networking technologies. However, the success of a networking technology depends not only on functional aspects, but even more on the cost of introducing it into existing infrastructures, with regard to end-system integration, cabling, interconnection and management. Hence, we will then investigate the evolution of networking infrastructures, to understand better the role different technologies can play.

4.1 Networking Requirements of Multimedia Applications

In this section we will analyze quantitative parameters like throughput and delay, and then focus on some functional requirements like point-to-multipoint capabilities and transmission reliability.

4.1.1   Throughput

Five years ago, many networking experts were convinced that multimedia communication would drive throughput requirements in the 1990s. Digital video transmission at 140 Mbps was considered a key networking application of the future. However, to reduce transmission bandwidth and storage requirements,

---

[*] Portions of this chapter are reprinted with permission from:
IEEE MultiMedia, Vol. 2, No.3 pp.42-59, Fall 1995, © 1995 IEEE

today's systems handle video data almost exclusively in compressed form. Two video compression standards (also see Chapter 3 of this book) are particularly relevant, namely, ISO MPEG [23] and ITU H.261 [24]. In terms of throughput they require 1.2 to 80 Mbps for MPEG and MPEG II and 64 kbps to 2 Mbps for H.261. Based on practical experiences, a total of 1.4 Mbps for audio and video together seems to be very attractive because it provides good video quality while allowing the use of commercial AV equipment (CD player) and transmission over T1-lines (1.5 Mbps). With regard to wide-area transmission cost, H.261 using 6*64 kbps, i.e., 384 kbps, is an attractive alternative. Existing H.261 implementations show that 64 kbps is only acceptable in almost static videos ("talking head video"), whereas the 384 kbps variant yields good results even for moving scenes. Hence, we conclude that current multimedia applications require a throughput of some 0.4 to 1.4 Mbps. This throughput is required for simplex, i.e., unidirectional streams because multimedia traffic often is of highly asymmetric nature. This must be kept in mind when designing bandwidth management schemes.

4.1.2   End-to-End Delay

Much harder to satisfy than the pure throughput requirements are the delay restrictions which interactive distributed multimedia applications impose on the communication channel. For example, think of the difficulties associated with a telephone conversation via satellite links. Round-trip transmission times around 0.6 seconds make it very difficult for the partners to have a normal conversation without prolonged silence or overlaps. Generally, it takes a few minutes to get both partners tuned to a high-level (mental) synchronization protocol. From this experience, we conclude that an end-to-end delay below 0.3 seconds is needed. Practical experiences with multimedia conferencing systems as well as the ITU standards suggests a maximum total end-to-end delay of up to 150 milliseconds for interactive video applications.

Traffic with an upper delay bound falls into the class of synchronous communication. Audio and video communication, however, is sensitive to more than the maximum transmission delay—it also expects a constant transmission delay for the different packets. This class of traffic is called isochronous communication. Hence, we distinguish the following traffic classes:

- *Asynchronous:* unrestricted transmission delay
- *Synchronous:* bounded transmission delay for each message
- *Isochronous:* constant transmission delay for each message

Isochrony does not necessarily have to be maintained across the entire path from source to sink, only at the final destination. Thus, isochrony can be recovered by a "playout buffer" at the sink. For practical purposes, synchronous communication with a sufficiently low delay bound can be used to implement an isochronous stream between source and sink. Therefore, we will not treat delay variance or delay jitter as separate items but rather confine our analysis to the upper delay limit. The drawback of recovering the isochronous characteristics by use of a

playout buffer is the introduction of additional delay. To meet the ITU requirements, we assume a maximum end-to-end delay of 150 ms. We can break this
down into at least four different contributing pieces:

- Source compression and packetization delay
- Network delay (access, transmission, and transit delay)
- End-system queuing and synchronization ("playout") delay
- Sink decompression, depacketization, and output delay

Video streams require the handling of 25 to 30 frames per second. Thus, real-
time compression/decompression times must not exceed 30 to 40 ms; they can, of
course, be smaller. Using another frame period for queuing and playout delay
leaves 60 ms for the maximum transmission delay.

Conversational multimedia applications like workstation conferences are
most useful across long distances; thus, we broaden our view from a single LAN
segment to a topology of interconnected local- and wide-area networks. Obviously, arbitrarily complex topologies will never be able to satisfy any given delay
requirement. However, assuming that a three-hop LAN-WAN-LAN route is a frequent topology and assuming further that the interworking units (gateways,
bridges, routers) also contribute some forwarding delay, we are then left with a
maximum acceptable delay on the order of 10 to 15 ms per hop. This is not a precise number but rather a rough approximation which can serve as a guideline for
the discussion of different networks.

4.1.3 *Multipoint Communication.*

Multimedia communication involves audio and video data, typical broadcast
information. Thus, integrated multimedia communication needs to support
multicast communication patterns in addition to the normal point-to-point
communication.

4.1.4 *Reliability.*

Traditional data communication strives to provide reliable end-to-end communication between two peers. Existing communication systems always use checksum and sequence numbering for error control and some form of negative or
missing positive acknowledgment with packet retransmission for error recovery.
If checksumming is not performed in hardware either at the MAC or link layer,
system performance can be badly impacted. The (negative) acknowledgment
with subsequent retransmission handshake adds more than a full, round-trip
delay to the transmission of this data. For time-critical data, the retransmitted
message may thus be useless altogether.

Therefore, the subnetwork itself should leave the proper error control and recovery scheme to the higher communication layers, which can provide the level of
reliability required, taking into account the impact on the delay characteristics.
A possible remedy for the conflicting goals of reliable and low-delay communica-

tions is the use of forward error correction (FEC) techniques; however, at this time, FEC techniques have not yet been implemented in any production network [1][7]. Chapter 5 provides a detailed discussion of reliability issues and FEC in the higher communication layers.

The issue of reliability becomes even more complex for multipoint communication. Currently, there exists no generally agreed-upon scheme for the semantics of a reliable multicast service. A more detailed discussion of reliable multicast issues follows in Chapter 6.

4.1.5 Channel Synchronization

When audio, video, and other data streams are coming from different sources via different routes, we need mechanisms to synchronize the different streams at the destination, to achieve the equivalent of lip synchronization. Synchronization can be achieved by a combination of time stamping and playout buffers. This is again an end-to-end function and not a characteristic of the specific network.

4.2 Networking Technologies

While the term "network" for typical local area networks clearly refers to the physical and the media access layers, it is less clear in the context of wide-area networks. In wide-area networking, services provided by carriers include a mixture of layers and functions: ATM provides a MAC-layer-like service; Frame Relay, a mixture of link and networking layer services; and X.25, a full-blown network layer service. To make matters worse, we find different networking layers being used one over the other, like IP over X.25. For simplicity, we focus in this discussion on the subnetwork issues, such as WAN networking services available from carriers, and ignore the problems created by constructing a logical network from various different subnetworks.

4.2.1 Relevant Parameters

To evaluate how well existing network technologies match the communication requirements derived above, we will use the following categories as a yardstick.

- **Throughput:** At least 1.4 Mbps per direction.
- **Transmission delay:** A maximum of approximately 10 to 15 ms. Here we do not treat delay jitter separately, because a playout buffering scheme can be used to achieve isochronous end-to-end behavior, and the respective additional delays have been accounted for in the last section.
- **Multipoint communication:** Available broadcasting and multicasting support. Obviously, multicasting is merely an addressing problem in broadcast type networks because all messages are distributed to all attached stations anyway. It is, however, more complex for switched networks.

- *Reliability:* Error control or recovery mechanisms integrated into the network. LANs support a hardware checksum mechanism and drop corrupted frames within their MAC layer. Given the inherent reliability of most of today's LANs, this is an acceptable strategy for data as well as for multimedia streams. Different strategies are mainly found in WANs, so we will cover reliability only for WAN technologies.

Because interstream synchronization is application specific, it is left to the end-systems and not treated within the network; we will ignore it here.

4.2.2 Ethernet

Ethernet at 10 Mbps is the most frequently deployed LAN technology. Assuming that some bandwidth needs to be left for data and control traffic and that Ethernets should not be loaded higher than 70% to 80% to keep collisions at an acceptable level, then only 5 to 6 Mbps are really available for multimedia streams. Thus, not more than four parallel compressed video streams can be supported. The drawback of the Carrier Sense Multiple Access with Collision Detection (CSMA/CD) access method is its nondeterministic behavior. In high-load situations, it exercises no control over access delay or available bandwidth per application. If a traditional application, such as remote file access, tries to utilize a large percentage of the available bandwidth, no mechanisms exist to ensure a fair distribution of bandwidth. In addition, Ethernet does not provide any access-priority mechanisms, and thus it cannot give preferred treatment to real-time traffic over conventional data.

Nevertheless, many of today's experimental multimedia applications use Ethernet as their transport mechanism, generally in a controlled and protected environment. With no more than three active stations on an Ethernet segment participating in a conferencing application, there is no real competition for bandwidth. In this kind of environment, Ethernet is perfectly suitable as a transport network.

Although in theory Ethernet can distinguish between up to 2^{46} different multicast addresses, in practice an Ethernet adapter manages a limited number of group (multicast) addresses.

Evaluation. Because of the missing delay guarantees, Ethernet is not a good network for distributed multimedia. However, it provides enough bandwidth for a few streams plus a multicast function, which makes it suitable for experimental setups with a limited number of stations.

4.2.3 100Base-T (Fast Ethernet)

100Base-T has been standardized by the IEEE 802.3 working group to scale Ethernet to 100 Mbps. It is frame-compatible with Ethernet and uses the same CSMA-CD access protocol; thus, it shares the same limitations with regard to

access delay characteristics. As with standard Ethernet, the maximum bandwidth utilization ranges from 50% to 90%, depending on configuration and frame
sizes. However, a recent survey [27] on commercially available 100Base-T adapters shows that throughput and access behavior are quite favorable, although not
deterministic, in smaller configurations. A drawback of the higher speed of a
100Base-T network is the relatively small maximum physical length, which is
250 m compared to 2500 m for a 10Base-T network. For video streams the limited frame size of 1,518 bytes is not appropriate. Given that the average frame
size of a 30 frames/s, 1.4 Mbps compressed video stream is close to 6 Kbytes, the
small frames lead to considerable segmentation overhead in end-systems and
routers.

Evaluation. 100Base-T provides enough throughput for a large number of multimedia streams. Delay guarantees cannot be given; in particular, any station on
the network can disrupt the multimedia stream through heavy traffic. Multicasting is available. In conclusion, 100Base-T is an attractive choice for small-to-
medium configurations, but it is not an ideal alternative for multimedia because
it relies on the underutilization of the capacity of a 100Base-T network and the
good behavior of all stations.

4.2.4   Isochronous Ethernet

A variant of Ethernet, called Isochronous Ethernet (Iso-Ethernet), has been proposed for the IEEE 802.9a Integrated Voice Data LAN (IVD LAN) standard. The
objective of 802.9a is to provide IEEE 802-like packetized MAC services together
with ISDN-like isochronous channels on a single, unshielded twisted pair (UTP)
of wires to a terminal. A first implementation of Iso-Ethernet was demonstrated
at the November 1992 COMDEX fair in Las Vegas. Five years later in 1997, only
a few commercial products are available.

   A more efficient coding technique frees up an additional 6 Mbps on the
standard Ethernet cable, providing 96 isochronous B-channels (64 kbps) along
with the standard 10 Mbps Ethernet channel. The extra 6 Mbps are used in a
slotted bus fashion; a synchronous frame at 8 kHz carries the B-channels. The
management and signalling of the B-channel is done via a separate signalling
channel (D-channel), using the ISDN Q.931 signalling protocol. One advantage
of this solution is that the existing 10 Mbps Ethernet cabling and adapters, as
well as previous investments in ISDN multimedia equipment, are largely preserved. Additional investments are mainly needed for the multiplexing unit that
combines and separates data and B-channels and for the central hub components. Further, the interworking of a local Iso-Ethernet and a wide-area ISDN
infrastructure is simple and efficient.

Evaluation: Iso-Ethernet is a shared-media approach with relatively limited
bandwidth and without multicasting support. It provides truly isochronous traffic, that is, optimal delay characteristics. Its ISDN-like channel structure is

designed for audio or H.261 coded video transmission, but it lacks bandwidth for MPEG coded streams. Thus, we can view it as a local-area ISDN extension on an existing networking base (Ethernet) rather than as a general solution to integrated multimedia communication.

4.2.5 Token Ring

The Token Ring access protocol is much better suited than Ethernet to support multimedia data. One reason is the advantage of 16 versus 10 Mbps of available bandwidth. More important is the availability of MAC-level priorities. They can be used to separate real-time data (high priority) and normal data (low priority).

Control and access to low-priority bandwidth remain unchanged. Stations can apply to a bandwidth manager for a portion of the reserved (high-priority) bandwidth. The bandwidth manager keeps track of the total allocated high-priority bandwidth to prevent overcommitment. To make the scheme work, an access control (scheduler or policing) mechanism is required per multimedia station, ensuring that each station does not exceed its allocated priority bandwidth.

Avoiding bandwidth overcommitment of the Token Ring does not necessarily guarantee acceptable access delays. If we consider a worst case scenario for a typical 1.4 Mbps video transmission, we can derive the following numbers:

- *Packet Size:* Large packets, although desirable for low system overhead, lead to increased packetization delays. Using 4-Kbyte packets in a 1.4 Mbps stream results in a 24 ms packetization delay, which already consumes most of the assumed 30 ms delay budget for compression plus packetization and depacketization and decompression. Hence, 4-Kbyte packets are the maximum allowable packet size in this case.
- *Packet Transmission Time:* A 4-Kbyte packet on a 16 Mbps ring is transmitted in 2 ms.
- *Token Forwarding Time:* Each Token Ring station contributes a very small time (less than 100 μs for a normal Token Ring) as a forwarding delay —negligible compared to the actual data transmission times, in particular, in the environment of large packets.
- *Maximum Access Delay:* The worst case delay will occur when one station transmits a large, low-priority packet and all other stations have a priority transmission request waiting. In the case of eight waiting stations (with a total of 8 × 1.4 Mbps = 11.2 Mbps allocated bandwidth), this situation implies a maximum access delay of 16 ms.
- *Total Transmission Delay:* Allowing 2 ms for transmission delay, the total delay from request to receipt is 18 ms. Although acceptable in a local environment, this delay can be slightly too large in an interconnected LAN/WAN/LAN scenario.

The situation improves slightly when the stations use small data packets. Packets of 64 or 128 bytes are common in transferring voice to keep packetization delays in the 10 ms range. A Token Ring can support over one hundred 64 kbps voice channels and stay below the required 10 ms delay boundary.

As in the above analysis, it is possible to derive algorithms that allocate bandwidth on the Token Ring and to calculate worst case delays [8]. The worst case delays can be provided to the bandwidth requester and can then be used to accumulate different delay components along a given network route. Such a scheme has been developed by Vogt and others [26].

Token Ring supports a addressing scheme similar to that of Ethernet; thus, a small number of multicast addresses can be managed by every station.

Evaluation: Token Ring provides enough bandwidth for a limited number of multimedia streams. Delay guarantees can be given provided priority access and a bandwidth management scheme are used. Multicasting is available, so 16 Mbps Token Ring is a viable entry network for multimedia communication.

4.2.6 Demand Priority

Another new 100 Mbps LAN technology is being standardized within the IEEE 802.12 working group as 100 Mbps Demand Priority LAN. Demand Priority was originally proposed as 100Base-VG, where VG stands for voice grade, and then put forward as 100Base-VG AnyLAN. It is an evolution of standard Ethernet and Token Ring to the 100 Mbps speed over Voice Grade (VG: category 3 UTP) cabling. Its main goal is increasing bandwidth while protecting existing wiring and interconnection investments. A more efficient coding scheme, together with four instead of one pair of wires, achieves a 10-fold speed increase with a modest increase in channel bandwidth at a cost similar to that of 10Base-T technology [32].

Demand Priority uses frame forwarding based on a round-robin access control scheme. Access control is managed by a central hub, relying on the standard star-wired LAN structure. Figure 4–1 shows operation of the access protocol. During an idle period, both hub and end-station send idle signals on two pairs of wires each. A station can signal a transmission request by putting a request signal on its upstream pair of wires (1). If the hub grants the request, it removes its idle signal from the downstream wires (2) to the sender and at the same time signals an incoming frame to all other stations (2), which forces them to lower their idle signal on the upstream wires (3). At this point, the requesting station sends the frame to the hub on all four wire pairs (3). The hub decodes the destination address of the incoming frame and forwards the frame to the destination station (4). At the same time, all other stations will receive idle signals (4), which allows them to raise their outgoing idle (5) and eventually signal a transmission request. When the source has completed sending the frame, it puts back the idle signal on the outgoing wires and silence on the incoming wires (5), thereby allowing the hub to restore the original idle state as soon as it has completed transmission of the frame (6).

The hub grants access to requesting stations on a round-robin basis. It can forward only one frame at a time, so it is a frame forwarding rather than a true

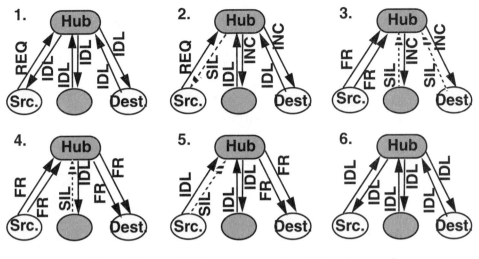

IDL = idle   REQ = request   INC = incoming
SIL = silence   FR = frame

Figure 4–1 Demand Priority Access Protocol

switching approach. Hubs operate in cut-through mode, which means that a hub begins to forward a frame as soon as the destination address has been decoded without waiting until the complete frame has been received. Hence, the forwarding delay introduced by the hub can be as low as a few microseconds. Delay bounds can be calculated and guaranteed for any given maximum packet size and number of attached stations. For instance, the transmission time for a 4-Kbyte packet is 0.33 ms. Thus, the access delay can be kept below 10 ms in a network of up to 30 stations. Demand priority supports a two-level priority scheme which can be used to favor multimedia over regular traffic and thus to reduce delays of real-time data.

Whereas the original 100Base-VG proposal supported Ethernet frames only, the later version, called 100VG-AnyLAN, supports both Ethernet and Token Ring frame formats. The same group addressing and multicast mechanisms as for Token Ring or Ethernet are available for Demand Priority LANs.

Evaluation. In terms of bandwidth, Demand Priority suits multimedia better than does 100Base-T because there is no restriction to low utilization for access delay reasons. Given the number of multimedia streams using high priority, access can be limited by a bandwidth manager, and the access delay can be guaranteed to be below the acceptable 10 ms. Multicasting is available, so Demand Priority is a viable alternative for multimedia communications, in particular, for topologies with a relatively small number of stations.

4.2.7   FDDI

The Fiber Distributed Data Interface (FDDI) protocol is an extension of the standard Token Ring protocol [31]. FDDI uses a timed token protocol. Every station measures the time that has elapsed since it previously received the token. When the station receives the token, it copies this measured current Token Rotation Time (TRT) into its Token Holding Timer (THT) and resets the TRT to a preconfigured Target Token Rotation Time (TTRT). Then, the station can send data until the THT expires. Figure 4–2 illustrates the FDDI protocol.

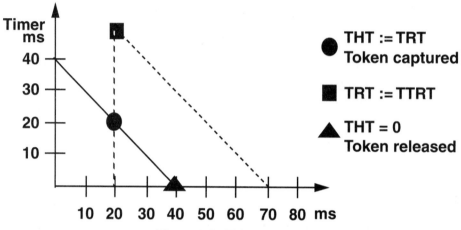

Figure 4–2 FDDI Timed Token Protocol

This scheme guarantees that each station will get access at least once within the preconfigured TTRT.

Conceptually the FDDI protocol is an extension of the Token Ring prototcol. Hence, the techniques to limit access delays, as discussed for the Token Ring, are directly applicable to FDDI as well. The larger bandwidth of 100 Mbps will obviously be able to support a larger number of multimedia stations.

In addition, FDDI also supports multimedia streams providing a synchronous traffic class. The operation of the synchronous traffic class is shown in Figure 4–3. Each station is assigned an interval of TTRT/n (with n being the number of synchronous stations on the ring) as guaranteed synchronous transmission bandwidth. When a station gets hold of the token, it can send traffic within the synchronous class for TTRT/n time units. After the synchronous time slot has been used up, the station can still transmit asynchronous traffic according to the normal timed token protocol as described above. The rotation time of the token is therefore increased to 2xTTRT.

This scheme allows synchronous, i.e., bounded delay traffic, with the delay limit (2xTTRT) configurable at ring initialization time. Unfortunately, a low delay bound leads to decreased ring utilization. In practical cases, the delay limit lies between 5 and 50 ms. Due to the increased bandwidth of FDDI, the transmission delay is largely governed by the access delay (transmission time for a 4-

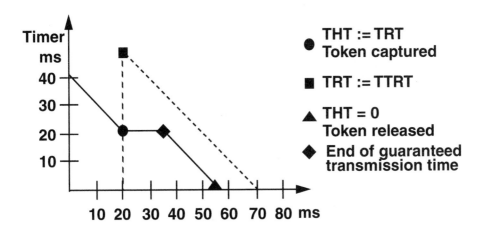

Figure 4–3 FDDI Synchronous Traffic Class

Kbyte packet is less then 0.5 ms). Note that although the synchronous traffic class has been built into the FDDI protocol, few existing FDDI implementations provide access to this traffic class.

FDDI supports multicasting in the standard broadcast LAN fashion.

Evaluation. FDDI supports multimedia communication well, thanks to high bandwidth, the synchronous traffic class, and available multicasting. When synchronous traffic is not supported by an adapter, a scheme similar to that for Token Ring, based on priority traffic and a bandwidth manager, can be used.

4.2.8 FDDI II

A network based on a slotted ring protocol known as FDDI II has been derived from FDDI to better support real-time traffic. Its isochronous capabilities are provided by using cyclic frames at an 8 kHz rate. The total bandwidth of 100 Mbps is split between up to 16 isochronous 6.144-Mbps wide-band channels (WBCs) and asynchronous FDDI bandwidth. Each frame starts with a template that describes the current allocation of channels and bandwidth. The distribution between asynchronous and isochronous bandwidth is dynamic. The management scheme needed to dynamically allocate the bandwidth between asynchronous and isochronous data is not yet defined.

Each WBC is composed of 96 isochronous 64 kbps ("B-") channels. This scheme provides for good interoperability with ISDN and Iso-Ethernet. In fact, FDDI II can be viewed as a backbone network for Iso-Ethernets.

Although FDDI II was derived from FDDI and claims compatibility with FDDI, this claim is only partially correct. Every FDDI II station can support normal FDDI traffic; however, a normal FDDI station cannot handle the FDDI II frame format. Thus, when only one FDDI station is on a ring, the advantages of FDDI II cannot be exploited. Because of this conflict, no migration path leads from FDDI to FDDI II, except reuse of the optical fiber. Thus, FDDI II is not a very strong contender right now because it is doubtful that there is enough room for a new, complex, high-speed network proposal with ATM already available.

Evaluation. FDDI II was designed to support constant bit-rate traffic. It provides truly isochronous channels with delays in the low millisecond range. Bandwidth is sufficient for a limited number of stations, and multicasting is available. Thus, it provides the ingredients for multimedia communication. At the same time, it has severe disadvantages in terms of complexity and compatibility.

4.2.9   DQDB

Another media access protocol that claims support for multimedia traffic is the Distributed Queue Dual Bus (DQDB) protocol [29][33]. It is most frequently used as access protocol for metropolitan area networks (MANs) providing Switched Multimegabit Data Service (SMDS) in the United States or the equivalent Connectionless Broadband Data Service (CBDS) in Europe. Note that neither SMDS nor CBDS relies on DQDB for its implementation; however, DQDB is frequently used for this purpose in the pre-ATM world.

Based on a dual slotted bus architecture, DQDB distinguishes between two kinds of slots: queue-arbitrated (QA) and prearbitrated (PA) slots. Access to a prearbitrated slot, or a fraction thereof, requires the respective station to reserve these bytes. Within the 125 μs bus cycle, each reserved byte or octet represents one 64 kbps isochronous channel. Thus, the PA mode is designed to support constant bit-rate isochronous traffic like ISDN channels. In current products or services, the reservation is still on a semipermanent basis; there is no dynamic allocation and deallocation of PA channels. Broadcasting or multicasting of isochronous channels requires the allocation of one upstream and a symmetrical downstream channel and is not available in current DQDB products.

Queue-arbitrated slots can carry connectionless and connection-oriented traffic. The connectionless access provides asynchronous IEEE 802 MAC-like services, including broadcast and three different priority levels. This part of the DQDB standard (IEEE 802.6) has been stable since 1990 and is supported in current DQDB products. It is used to implement SMDS or CBDS services. It does not provide any throughput or delay guarantees.

Currently, the IEEE 802.6 working group is defining connection-oriented services with throughput and delay guarantees. These services are modeled closely after the ATM AAL 3/4 and AAL 5 services (see section 4.2.14) to maximize interoperability between both networks. Figure 4-4 shows the different DQDB functional blocks. A stream-oriented convergence function with additional support for multimedia streams is also under discussion but not yet defined. All three connection-oriented services use a Guaranteed Bandwidth (GBW) algorithm to ensure throughput and delay guarantees. The GBW includes reservation and access control mechanisms. However, the draft standards for the three connection-oriented services and their corresponding convergence functions have

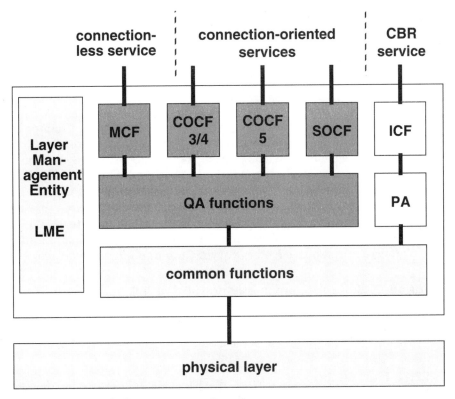

MCF:	MAC Convergence function
COCF 3/4:	Connection-Oriented Convergence Function using AAL 3/4
COCF 5:	Connection-Oriented Convergence Function using AAL 5
SOCF:	Stream-Oriented Convergence Function
ICF:	Isochronous Service Convergence Function
QA:	Queue-Arbitrated Access Function
PA:	Prearbitrated Access Function

Figure 4–4 DQDB Functional Blocks

yet to be approved, so it unclear when and whether there will be products implementing these services.

Evaluation. DQDB in its future version seems conceptually well suited to support multimedia in terms of delay characteristics. However, current metropolitan area DQDB-based SMDS or CBDS services lack these advantages. They are restricted to the asynchronous traffic class and a few static CBR channels.

Although 150 Mbps is certainly a lot of bandwidth in the local area, it is too little to transport a large number of video streams in a metropolitan area network. Also, multicasting is difficult. Hence, DQDB in today's form targets asynchronous traffic rather than multimedia. Today, the connectionless asynchronous service can be used for experimental multimedia configurations. Delay behavior will be acceptable as long as the MAN is lightly loaded, which is frequently the case in current public MANs. Whether the more suitable connection-oriented, guaranteed-bandwidth services will become available in the future is doubtful.

4.2.10 X.25 Packet-Switching

X.25 is a network layer service and is covered in more detail in Chapter 5; it is, however, one of the most common point-to-point wide-area services used in Europe. Therefore, we will here look at X.25 from a subnetwork point of view.

X.25 has been designed to provide reliable data transport over relatively low-speed and unreliable links. Its processing requirements and window mechanisms make X.25 suitable only up to 2 Mbit/s. It does not provide any mechanisms to plan network utilization and avoid congestion; instead, its sliding window-based, per-hop flow control and error recovery mechanism handle congestion as it occurs and hence lead to varying and unpredictable delays.

Evaluation. X.25 services are mostly provided at link speeds up to 64 Kbit/s. Neither transmission delay nor its variance are predictable. In addition multicasting is not available for X.25 services. Thus, X.25 does not suit multimedia networking.

4.2.11 Frame Relay

During the 1980s, faster and more reliable wide-area links became available. In this environment, X.25 was not a suitable protocol because of the relatively high processing overhead in each intermediate switch. On the other hand, with reliable physical connections, the need for building reliability into the lower-layer communication protocols disappears, and end-to-end error control becomes completely sufficient. Frame relay [33] was developed to overcome the limitations of X.25 in a high-speed environment. It uses the LAP-D frame structure with a 19-bit data link connection identifier (DLCI) for routing. A 2-byte frame checksum (FCS) is used to detect corrupted frames. This FCS algorithm limits the frame size to a maximum of 4 Kbytes. However, not all FR equipment supports 4-Kbyte frames; the lowest maximum frame size required by the standard is 1600 bytes, sufficient to transport Ethernet frames without segmentation.

The frame relay protocol itself does not provide error recovery or flow control mechanisms. Corrupted frames are discarded and not recovered. The frame header provides two bits to notify the end-system of congestion situations discovered along the path. A frame relay switch can set these Forward and Backward Explicit Congestion Notification (FECN and BECN) bits in case it experiences any congestion. Upon receipt of a FECN or BECN, the user decides whether to throttle his or her traffic. Thus, no built-in protocol mechanisms negatively impact the transfer delay characteristics. In a multimedia environment, the

BECN flag can be used to reduce the sending rate (also see media scaling in Chapter 5 and [13]).

On the other hand, the switch itself introduces latency that depends on the current load, the line speeds, and the frame size. Whereas packetization delays on a 2 Mbps link for a 2-Kbyte packet are significant (8ms), such delays are negligible on higher speed (E3, T3) lines (1 ms for 4 KByte on 34 Mbps).

Frame relay provides a rudimentary priority scheme. The Discard Eligibility (DE) bit indicates frames that the switch may discard in case of congestion. There are two ways of using the DE bit. First, the user can set the DE bit for less important messages, which is a useful mechanism to support media scaling. Second, when a frame relay virtual circuit (VC) is set up, a bandwidth value called the Committed Information Rate (CIR) is assigned to it. When the interface speed exceeds the CIR, the user may put more load on the network than requested, but then all excess frames will be tagged with the DE bit and may be discarded by any switch in its path in case congestion occurs. Thus, frame relay provides a useful intrastream—but no interstream—priority scheme. The delay introduced by the FR switch itself depends on the load that is generated by other links.

Frame relay was designed for wide-area backbones interconnecting LANs or communication controllers. In practice, frame relay is being used in two different ways.

First, and most frequently, frame relay is used as a framing protocol over leased-line services. In this environment, FR does not introduce any uncontrollable delays into multimedia streams. The only disadvantage here is the lack of a suitable interstream priority mechanism. The packetization delays mentioned above are not a protocol function but merely an indication of insufficient throughput.

Second, frame relay is, of course, used to build switched networks. Current frame relay services are mostly based on preconfigured permanent virtual circuits (PVCs). There are plans to provide special multicast VCs based on these PVCs, but it is questionable whether this static multicasting approach is a sufficient base for the distribution of multimedia information. Given the focus of using frame relay for LAN backbone purposes, the market demand to provide dynamically switched virtual circuits (SVCs) is relatively low.

Evaluation: Frame relay is a layer-two protocol, with services provided on different physical networks from 1.5 to 45 Mbps. Once the wide-area bandwidth is physically available, frame relay can transport time-critical data like audio or video. The protocol does not add substantially to the delays imposed by the line itself, but frame relay switches may add unpredictable delays. Multicasting is planned but not yet available. As a framing protocol on a link connecting two LANs, frame relay supports real-time streams. On the other hand, switched frame relay services do not satisfy the stated delay and multicast requirements.

4.2.12 IP Packet-Switching Networks

Packet switching networks based on the Internet protocol (IP) generally consist of a variety of different subnetworks of various technologies. An IP packet commonly traverses LANs, X.25 links, leased lines, and frame relay links on its way from source to destination. In this section, we focus on the impact of the basic IP packet switching mechanisms on multimedia communication. A broader discussion of IP-related issues like routing policies is deferred to the next chapter.

IP imposes no bandwidth restriction; thus, it will deliver whatever the link service can provide (minus a small fraction of overhead). Since IP is a connectionless network layer, different packets may traverse various routes between source and destination, encountering greatly varying link and router delays. Further, IP does not provide any mechanisms for congestion control, which again can lead to varying delays and packet losses. These observations are valid for the current version 4 of IP. They will be corrected by the introduction of the flow concept in the future version of IP, called IP next generation (IPng) or, more precisely, IP version 6, which is discussed in detail in Chapter 5. The IPv6 flow concept also allows to distinguish high-priority real-time packets from normal data and to identify different real-time streams. In fact, flows introduce a soft connection model into IP.

The IP multicast extensions [12] utilize the LAN broadcast mechanism to distribute IP packets to all members of an IP multicast group on a LAN segment and packet replication on nonbroadcast networks like point-to-point wide-area links. With this approach, IP provides an unreliable 1-to-N multicast service, which is the basis of the multicast backbone (MBone) and its associated multimedia applications (for a more detailed discussion, see Chapter 5). Management of the different multicast address groups is handled by the *Internet Group Management Protocol* (IGMP), which runs directly on top of IP.

Evaluation. As a networking layer protocol, IP does not preclude the efficient use of high-speed communication links. No delay or delay jitter guarantees exist because of its connectionless characteristics. Multicasting is available, so despite its unpredictable delay behavior, IP is being used as an experimental multimedia communication platform.

In practice, the Internet is the most-heavily used, wide-area, multimedia network. The poor quality of audio and video transmissions is not surprising, given the general lack of wide-area bandwidth. In addition, varying delays occur for the reasons explained above. However, due to the Internet's universal availability, IP has great merits for promoting the development of distributed multimedia applications.

4.2.13 ISDN

Looking at the wide area, the Integrated Services Digital Network (ISDN) was designed to support a large variety of different services, from data over voice to fax and video. It provides circuit-switched, synchronous, 64-kbps channels, which can be used for continuous bit-stream-oriented traffic (CBO or CBR), such as H.261 coded video, or for packetized communication.

ISDN is widely available in Europe. The basic rate interface (BRI) provides 2 x 64 kbps of bandwidth. The primary rate interface (PRI) with its 24 (US) or 30 (Europe) 64-kbps channels is the solution for higher throughput requirements. However, depending on the internals of the ISDN network, different channels may use different internal routes; thus, bundling several channels to form a larger channel may require buffering at the receiver to compensate for delay shifts.

When ISDN is used for CBO, end-to-end traffic delay is constant and negligible. However, in the environment of interconnected LANs/WANs, we also must consider packetized multimedia data because today's LANs are not able to handle CBO traffic. In this case, the low throughput introduces relatively high packetization delays. A simple calculation (see Table 4–1) shows these delays.

Bandwidth				
Packet Size	64 kbps	128 kbps	384 kbps	1920 kbps
256 byte	32 ms	16 ms	5 ms	1 ms
1 Kbyte	128 ms	64 ms	21 ms	4.3 ms
4 Kbyte	500 ms	250 ms	85 ms	17 ms

Table 4–1 ISDN Packetization Delays

The guidelines derived above in section 4.1.2 restrict the packet size to 1 Kbyte for 6 bundled B-channels (384 kbps) and 4 Kbytes for 30 bundled B-channels (1920 kbps). Basic rate connections require even lower packet sizes (128 or 256 bytes) to keep the delay at an acceptable level, which can lead to considerable performance problems in the attached end-systems and intermediate gateways. Note that the packetization delay and the ISDN transmission delay are fixed; thus, no additional playout buffering is required.

Despite these throughput and delay restrictions, there are several advantages of ISDN over other wide-area networks for multimedia communication:

• Wide availability
• Bandwidth scalability at the primary rate interface, in steps of 64 kbps
• Isochronous characteristic
• Support for CBO as well as packetized traffic

For nonpermanent usage, i.e., for sessions of limited duration, ISDN also provides attractive tariffs when compared to leased lines.

ISDN does not support any multicast function within the network. However, to provide point-to-multipoint connections, in particular for conference applications, equipment external to the ISDN, called Multicast Control Units

(MCUs), are being used. Each station participating in an ISDN conference session builds a point-to-point connection to the MCU. The MCU then mixes all incoming audio channels and distributes the combined signal over point-to-point links to all participating stations.Whereas the concept of mixing is useful for audio streams, it is not directly applicable to video or other data streams. Hence, when using an MCU for video or data, the application has to select one incoming stream for distribution rather than mixing several incoming streams into one outgoing stream.

Evaluation: ISDN provides limited bandwidth up to 2 Mbps with isochronous characteristics. Since neither X.25 nor frame relay services satisfy our requirements, ISDN is currently the only widely available choice for interactive, wide-area, multimedia communications, aside from leased line services. The lack of built-in multicast services requires the use of special equipment (MCUs) to set up multipoint conferences or distribution services.

4.2.14 ATM

Asynchronous Transfer Mode (ATM) is a cell-based multiplexing and switching technique around which new network architectures are being built. The International Telecommunication Union (ITU, previously called CCITT) defined ATM as the base technology for the future public B-ISDN network [14][33] in 1988.
Thus, ATM is the strategic networking technology for wide-area carriers. Progress in the development of ATM standards and technology was relatively slow until, in 1991, the ATM Forum, founded by equipment manufacturers,

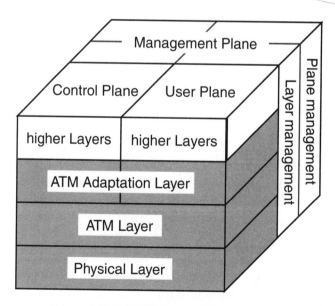

Figure 4–5 B-ISDN: Protocol Reference Model

focused on the definition of private ATM networks for the local- and wide-area. In the meantime, the ATM Forum with its over 800 members, including almost all major providers of data and telecommunication equipment, and the ITU cooperate in the ATM standardization, so that both sets of specifications are sufficiently similar. The ITU has defined the protocol reference model for B-ISDN shown in Figure 4–5.

The Physical Layer

The lowest or physical layer defines the mapping of ATM cells to the physical media and the parameters of the physical transmission. The CCITT initially targeted a 155-Mbps optical transmission. The ATM Forum first defined the use of the 100 Mbps physical FDDI transmission standard for local area ATM networks. Other options include unshielded twisted pair (UTP-3) cabling at 25.6 155-Mbps or 51 155-Mbps, or synchronous lines such as T3 (45 Mbps) or E3 (34 Mbps) at the user network interface (UNI). The list of supported physical layers, as shown in Table 4–2, is still growing.

rate	name and framing	defined by
622 Mbit/s	SDH STM4/SONET STS-12 over SMF	ITU
155 Mbit/s	SDH STM-1/SONET STS-3C over SMF, MMF, STP, UTP-5	ITU, ATM Forum
155 Mbit/s	cell based over MMF, STP, UTP-5	ATM Forum
100 Mbit/s	cell based MMF (TAXI)	ATM Forum
51 Mbit/s	UTP-3, MMF, SMF	ATM Forum
45 Mbit/s	G.804/T3	ATM Forum/ANSI
34 Mbit/s	G.804/E3	ATM Forum/ETSI
25.6 Mbit/s	STP, UTP-3, UTP-5	ATM Forum
2 Mbit/s	E1	ATM Forum/ETSI
1.5 Mbit/s	T1	ATM Forum/ANSI

SDH: synchronous digital hierarchy (ITU)
STM: synchronous transfer module (ITU)
SONET: synchronous optical network (ANSI)
SMF: single-mode Fiber
MMF: multi-mode Fiber
TAXI: transmitter/receiver interface- physical FDDI interface
STP: shielded twisted pair
UTP: unshielded twisted pair

Table 4–2 ATM Physical Layer Options

Besides the line speed and the form of transmission (optical vs. electrical), we also distinguish the framing schemes used. In the local area, we find pure cell-based mappings like the 100 Mbps TAXI or the 25.6-Mbps interfaces. The cell-based interfaces either use continuous streams of cells, where ATM layer cells are interleaved with physical layer cells, which can be used for operation and maintenance (OAM) of the physical transmission medium, or they send cells asynchronously, separated by idle signals.

Although they were originally designed for the wide-area, we now find more and more framed interfaces for the local area. In frame-based transmission, ATM cells are transported as payload of a larger (SDH/SONET or PDH) transmission frame. The transmission frames can carry OAM information within special fields reserved in each frame, and other non-ATM information, together with ATM cells within each frame. Framing increases the flexibility and manageability of the transmission system as well as the transmission overhead and implementation complexity. However, as SDH/SONET framing is available in VLSI today, the additional price is outweighed by the economic advantages of using a single transmission infrastructure for the local- and the wide-area. The only exception is the low-end bandwidth range (25.6 Mbps and below), where bandwidth consumed is too precious.

In terms of throughput even 25 Mbps of dedicated bandwidth exceed the 1.4 Mbps required for audio and video streams by far. Currently, we see the first ATM interfaces at 622 Mbps, and even 2.4 Gbps are not too far away.

The ATM Layer

The ATM layer is a switching and multiplexing layer. It defines the structure of the ATM cells. Cells are information containers of 53 octets: 5 octets of header information and 48 octets payload. This cell format is completely independent of the underlying physical layer, which is one of the great strengths of the ATM architecture.

ATM uses a label swapping concept similar to that of X.25 for routing cells from one switch to the next. Routing decisions are based on virtual circuit identifiers (VCs) and virtual path identifiers (VPIs). Thus, ATM is a connection-oriented network. The switch control function defines the mapping of network addresses to VPI/VCI pairs during the connection establishment phase.

Because of the small size of the ATM cells, the high speed of the transmission links, and the switching speed of the nodes, ATM provides very low latencies. The small ATM cells can be interleaved with a very fine granularity, which reduces segmentation and buffering delays. Different buffering techniques and buffer dimensions for ATM switches to minimize cell loss and switching delays are under discussion.

Despite much ongoing research in ATM traffic management and switch dimensioning [19], there is little practical experience with large and heavily loaded ATM networks, and therefore the validation of the various traffic models is very difficult.

The ATM Adaptation Layer (AAL)

For most applications, the cell-based ATM layer is not an appropriate interface. CBR applications require a bit-stream interface, and normal data applications are not communicating in 48-byte small messages. Thus, the ATM adaptation layer has been designed to bridge the gap between the cell-oriented services of the ATM layer and the application requirements. The ITU defines the four different service classes for B-ISDN as shown in Table 4–3.

	Class A	Class B	Class C	Class D
Timing Relation Between Source and Destination	required		not required	
Bit Rate	constant	variable		
Connection Mode	Connection-oriented			Connection-less

Table 4–3 ATM Service Classes

Class A services target constant rate synchronous bit streams, from ISDN to T3 links. Class B targets variable rate compressed audio or video streams. Class C and Class D model existing data communication services.To support these different classes of service, different adaptation layers have been defined:

- *AAL Type 1* for class A services provides packetization/depacketization functions at the UNI to convert CBR into cell-based traffic, and vice versa. At the receiving side, a fully synchronized (clocked) bit stream has to be delivered, which requires tight delay control within the network. AAL 1 is a prerequisite for ATM/ISDN interworking, as well as for synchronous voice and H.261 coded video transmission. Therefore, it is already available on early wide-area ATM products.

- *AAL Type 2* supports class B services. Implementing AAL 2 is difficult. The variable bit rate makes it difficult to reserve resources for this kind of traffic. Either the peak bandwidth is reserved—not very efficient—or delay and loss guarantees may not be held. Currently the ITU is defining a first version of AAL 2 with a slightly different focus. It is targeted particularly at aggregating multiple variable-rate compressed voice streams into one ATM virtual circuit ("trunking"), which is one important class B service for wide-area carriers. However, products are not expected to be available in the near future.

- **AAL Type 3/4** implements class C and D services. As ATM is inherently connection-oriented, connectionless service needs to be provided by connectionless servers (CLS), which themselves are accessed through connection-oriented communication. Thus, there is no need to distinguish service classes C and D at this level. This fact was only discovered later during the standardization process, and thus the earlier types 3 and 4 are now joined into a common AAL 3/4.

  The main functions of the AAL 3/4 are segmentation and reassembly of messages to cells, and vice versa. In addition, AAL 3/4 provides a message identifier (MID) field, allowing an interleaved transport of different messages over the same VC. This feature is useful in the context of connectionless or multicast servers. It allows the server to forward individual cells of a message without reassembly at the server.

- **AAL Type 5** as defined by the ATM Forum also provides class C and D services. AAL 5 was initially called the Simple and Efficient Adaptation layer (SEAL). It does not support message interleaving; thus, there is no need for a MID field per cell. Consequently, AAL 5 provides for better utilization of the available bandwidth. Each 53 Octet cell of AAL 3/4 carries a 5-byte cell header and an additional 4 bytes of segmentation overhead. In AAL 5, the 4-byte per cell segmentation overhead can be reduced to 6 bytes per message, not per cell (2-byte length field plus 4-byte CRC). For larger messages, only a fraction of a byte per cell is used as segmentation overhead, and the cell overhead is reduced from 17% to 11% for large packets. The reassembly process itself is also slightly simplified since each cell requires less checking.

  One drawback is that a corrupted cell will always lead to a discarded message in AAL 5, whereas AAL 3/4 provides means to localize bit errors to individual cells. This feature is interesting in the context of multimedia communication, since some media streams might be able to use partially correct messages.

At this time, there is no clear direction for a further AAL. There are active discussions within the Audiovisual Multimedia Services (AMS) working group of the ATM Forum on an AAL suitable for the support of packetized multimedia streams, in particular for MPEG II coded videos. Issues under discussion are the use of forward error correction (FEC) techniques to increase link security to a level where no extra error recovery will be needed. Another issue is the support of synchronization requirements of MPEG II. However, likely we will see optional extensions to AAL 5, rather than a completely new AAL 6. One extensions is a convergence layers above AAL 5 supporting MPEG transport streams.

Currently, the trend goes more and more towards AAL 1 for interworking with ISDN and other synchronous services, and AAL 5 for almost everything else. From a multimedia perspective, the advantages of AAL 3/4 are limited. The error handling mechanisms of AAL 3/4 seem to be useful, whereas the multiplexing capability has little added value, assuming there are sufficient number of VCIs available in the first place. Thus, efficiency and availability are probably more important than added function. Although initially not anticipated by the ITU, there is a demand for using AAL 3/4 or AAL 5 over bounded-delay services in order to support packetized interactive multimedia traffic. In early pilot ATM networks, which tend to be small and lightly loaded, this is not a problem because of the lack of contention. In future ATM networks enhanced traffic management capabilities will be needed to guarantee the required throughput and delay characteristics.

AAL services and protocols are implemented within the ATM end-systems, completely transparent to the ATM switches, except for the signalling plane. Hence, we may see new AALs or AAL extensions evolve as new application requirements are identified, withou major impact on the network elements, i.e. the switches.

Signalling

Connection management in ATM networks follows the out-of-band scheme known from the ISDN protocol architecture. A signalling protocol (Q.2931; previously Q.93B) allows end-systems to communicate their connection management requests to the switch control. The switch control can then set up the proper route and allocate the required resources. It then communicates the connection request to the next node or the target end-system. The widely available version (ATM Forum UNI Vs. 3.1 [4]) of the ATM signalling protocol allows the establishment of bidirectional point-to-point and unidirectional point-to-multipoint connections. On point-to-multipoint virtual circuits, endpoints can be dynamically added or dropped. However, today it is always the sender node that can add a new node to the multicast tree. The new UNI Vs. 4 [5] will provide leaf-initiated joining to support receiver-initiated multicast—a function that is extremely useful for modeling the typical behavior of a broadcast media receiver. Listeners can join and leave broadcast or multicast transmissions without the sender being aware of it.

The ATM connection establishment request also contains data structures to describe the characteristics of the traffic to be expected on this connection. Traffic specification and management are described in the next section. With regard to the signalling of quality of service parameters, UNI 3.1 signalling supports only a "take it or leave it" approach; negotiation of these parameters is a function of the UNI version 4.0.

Traffic Management

The model behind ATM traffic management is a contract that a user agrees on with the network. User and the admission control function of the network agree on a traffic and a quality of service (QoS) description.

The ATM traffic descriptor is based on the following characteristics of the cell stream:

- **PCR**: Peak Cell Rate (cells/s)
- **SCR**: Sustainable Cell Rate (cells/s)
- **MBS:** Maximum Burst Size (cells), also specified as
- **BT:** Burst Tolerance= (MBS-1)/(1/SCR-1/PCR)
- **MCR:** Minimum Cell Rate (only for ABR traffic)

It is the responsibility of the user to make sure that he does not exceed the agreed-upon traffic characteristics. The usage parameter control (UPC) or policing function of the network controls the incoming cell stream and may discard any nonconforming cells.

This traffic description based on cell rates may be suitable to allow the admission control to do the calculations necessary to gain statistical multiplexing advantages by combining several streams without exceeding the line or switch capacities, as well as guaranteeing low buffering delays. It is, however, very difficult for a user or an application to accurately provide these parameters since there is no direct relationship between a video frame rate or a video resolution to the cell rate characteristics of a compressed video stream.

For the specified traffic characteristics, network and user agree on a certain quality of service (QoS), which the network provides for the acceptable input traffic. The QoS parameters used are

- **CLR**: Cell Loss Ratio (number of lost cells / number of transmitted cells)
- **CTD**: Cell Transfer Delay (network entry to exit delay)
- **CDV**: Cell Delay Variation (CTD variance)

Depending on which traffic descriptor parameters are specified, ATM traffic can be classified into five different service categories or traffic classes, listed below and summarized in Table 4–4.

- **CBR**: Continuous bit rate traffic with fixed delay and cell rate for synchronous services like ISDN emulation or synchronous audio/video channels
- **Real-time VBR**: Variable bit rate traffic for compressed audio and video data streams with real-time characteristics (CDV specified)
- **Non-real-time VBR**: Variable bit rate traffic for compressed audio and video data streams with real time characteristics (without CDV specified)
- **ABR:** Available bit rate traffic for data communication with negligible cell loss
- **UBR**: Unspecified bit rate for traffic without known characteristic

From a multimedia point of view, mainly CBR and VBR are of interest, CBR being suitable for any synchronous, constant bit-rate stream, and VBR for variable rate compressed real-time stream. However, the provision of good rt-VBR traffic management with acceptable performance guarantees is an expensive undertaking in terms of network resources.

	CBR	rt-/nrt-VBR	ABR	UBR
CLR	specified		specified	unspecified
CTD	specified		unspecified	
CDV	specified	optional	n.a.	
PCR	specified			
SCR/BT	n.a.	specified	n.a.	
MCR	n.a		specified	n.a.

Table 4–4 ATM Traffic Classes

Evaluation: Although the ATM protocol architecture provides the ingredients required to support the requirements of multimedia applications, much of the architecture is not yet implemented. The first wave of local area ATM products focused on "best effort" data communication, using LAN emulation or RFC 1577 [22][3], which actually hide QoS control from the user. Although many of the traffic management issues remain unsolved, ATM networks provide ample bandwidth for multimedia streams. Delay and delay jitter are sufficiently small under normal load. Delay control and reliability will have to be analyzed under heavy load condition, and suitable traffic management is yet to be developed. Beyond that, ATM does provide for multicast communication, so it is the best match for the multimedia requirements listed above.

4.2.15 IP-Switching

Yet another new networking technology, collectively referred to as IP-switching, appeared during the last year. Rather than being a single switching technology, it is a collection of approaches that share a common idea, namely, to develop a network architecture that combines the speed and efficiency of ATM switching with the flexibility of IP routing. Currently, several proposals are on the table, and the IETF has established a working group called MPLS (Multiprotocol Label Switching) to define a standardized approach. At this time, only one of the architectures, namely, Ipsilon's IP-Switching (see Figure 4–6), is supported by commercially available products.

The basic idea is similar in all the different proposals, so we will explain the principles of this technology using Ipsilon's approach [20].

Each IP switch, sometimes called "integrated switch/router" or "cell switched router," consists of two components: a cell switch and an IP router, sometimes also called IP switch controller (IPSC). The operation of IP-Switching is demonstrated in Figure 4–7.

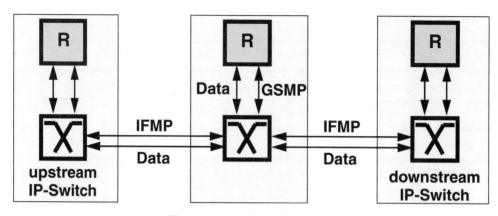

Figure 4–6 Ipsilon's IP-switching

We can best describe this operation in six steps:

1. Initially, all IP-hosts and IPSCs are connected via default VCCs, and data is forwarded through the IPSCs (*store and forward*).

2. An IPSC analyzes the stream of IP-packets and identifies an IP-flow.

3. Once a flow has been identified, the switch controller establishes a direct VCC to the upstream IP node (using GSMP).

4. The downstream-IPSC does the same.

5. The switch controller informs the upstream node through an IFMP message to use the new VCC.

6. The switch controller instructs the ATM-switch to connect the respective VCCs directly (port mapping).

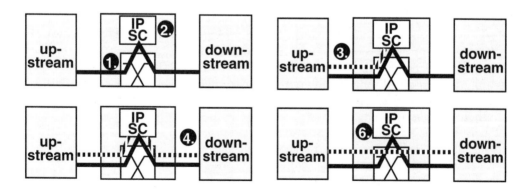

Figure 4–7 IP Switch Operation

From then on, all packets of the same flow are switched directly via the port mapping feature through the network. Whereas the protocols to exchange flow information (IFMP [16]) and to control the switches (GSMP [28]) are published as informational RFCs, the real art, namely, the algorithms to detect IP flows, is proprietary technology and will distinguish IP-Switching products of different vendors. Ipsilon claims that more than 80% of all packets and 90% of all bytes belong to detectable IP flows. Since the architecture is built around individual flows, it is in principle possible to provide per flow QoS with IP Switching, It will also be easy to incorporate and take advantage of the IPv6 flow labels into the flow identification scheme. The scheme also extends to multicast traffic. Because the IPSCs are IP routers, the standard IGMP protocol can be used to distribute group membership information through the network. Replication of multicast packets can be done efficiently by providing a 1-to-n port mapping at the switch level. At the switch level there are no major performance differences between standard ATM switches and IP switches. Therefore, the suitability for multimedia communication is at the same level as with standard ATM.

Besides the flow-oriented scheme implemented by Ipsilon, two topology-oriented schemes, called *"Tag Switching"* and *"Aggregate Route Based Switching (ARIS)"*, have been proposed by Cisco [30] and IBM, respectively. These architectures do flow identification based on routing information and can aggregate application level flows into "network route flows." Thereby, they can utilize VCs more efficiently than can the original IP Switching; however, aggregation of streams is not an advantage when one is trying to implement per (application) flow QoS control.

4.2.16 Summary of Network Characteristics

Table 4–5 summarizes the above evaluation. In addition, it shows, wherever possible, some numerical values for delay bounds and provides a separate column to indicate commercial availability of the respective technology.

From this summary we conclude that although there are several network technologies that promise good support for multimedia traffic, not all are actually available today. Of those that match the outlined multimedia requirements, only ATM is actually widely commercially available.

Yet another aspect has not been discussed so far, namely LAN/WAN-integration. We can group the technologies according to their multimedia model—support for CBO vs. packetized multimedia traffic—in the following way:

- CBO support: Iso-Ethernet, FDDI II, DQDB (PA), ISDN, ATM (AAL 1)
- Support for packetized multimedia: Token Ring, Demand Priority, FDDI, ISDN, ATM (AAL 3/4, 5), IP-Switching

Today the CBO group can rely only on ISDN WAN and ATM AAL 1 support, whereas the packetized group can build on both local- and wide-area networks. In addition, no CBO support exists on Ethernet, Token Ring, or FDDI, which together constitute the vast majority of existing LANs. The one strong argument in favor of a CBO-based environment is the smooth integration of computer and traditional telecommunication services. The basis for such an integration is the 64 kbps isochronous ISDN channel, around which the H.320/H.261 series of standards for real-time audio/video compression have been defined. However, the total lack of isochronous 64 kpbs channel support in the existing computer network infrastructure severely hampers the proliferation of this approach.

Network	Band-width (Mbit/s)	Dedicated vs. Shared	Transmission Delay	Delay Variance	Broad-cast	Avail-able
Ethernet	10	shared	random	•	+	+
Iso-Ethernet (isochr.part)	10+6	shared	fixed < 1 ms	0	-	+
Token Ring	4/16	shared	configuration dependent < 20 ms	max.	+	+
100 Base-T	100	shared	random	•	+	+
Demand Priority	100		configuration dependent < 10 ms	max.	+	+
FDDI	2*100	shared	configuration dependent < 20 ms	max.	+	+
FDDI II (isochr. part))	n*6	shared	fixed < 1 ms	0	+	-
DQDB QA Acc.	2...155	shared	random	•	+	+
DQDB PA Acc.	2...155	shared	fixed	0	-	-
X.25	< 2	dedicated	random	•	-	+
Frame Relay	< 50	dedicated	random	•	(+)	+
IP	unlim.	dedicated	random	•	+	+
ISDN	n*0.064	dedicated	fixed < 10 ms	0	-	+
ATM	25..2048	dedicated	bounded < 10 ms	max.	(-)	+
IP Switching	...2048	dedicated	bounded < 10 ms	max.	(-)	+

delay variance:
• = asynchronous network without delay jitter control
max.= synchronous network with delay variance between 0 and max. delay
0 = isochronous network with constant delay

Table 4–5 Survey of Network Characteristics

4.3 Networking Infrastructure Evolution

Over the last 20 years, networking technology has advanced dramatically, particularly for wide-area technologies, as shown in Figure 4–8.

Whereas in the past a distinct difference existed between local and wide-area networks in terms of throughput, physical extension, and reliability, this gap is closing now, at least with regard to technical capabilities. Users do not want to deal with LAN/WAN boundaries; they want networking services independent of physical or organizational boundaries. This is particularly true for multimedia applications like workstation conferencing or video distribution, which are most useful when available beyond the limits of a single LAN. Seamless LAN/WAN integration is an important consideration for future network evolution.

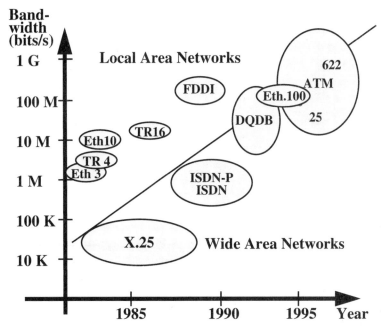

Figure 4–8 Network Bandwidth Evolution

4.3.1 Network Deployment Issues

The introduction of new networking technologies is an evolutionary rather than revolutionary process. Massive investments of wide-area carriers and local-area users cannot be replaced easily and must be protected. The main investments are in:

- Cabling
- End-systems and adapters

- Switching and interconnection units
- Network management and operation
- Existing networked services and applications

Further operating expenses like wide-area traffic charges greatly impact the evolutionary process. Thus, even political decisions like telecommunication liberalization strategies influence direction and speed of the networking evolution.

4.3.2   WAN Evolution

In the wide area, cabling consumes the bulk of networking investments, with the local loop absorbing the largest fraction of these costs. A second major investment goes to the network carrier nodes, from switches to satellites. Therefore, it is mandatory to use the components of the existing WAN infrastructures to their limits and to deploy new infrastructures in an evolutionary manner. This guarantees undisturbed operation of existing services, as well as the provision of advanced services with a growth path into the foreseeable future.

Today there are five main wide area networking infrastructure components:

- Public and private packet switched data networks based on X.25 or Frame Relay.
- The public telephone infrastructure and its associated telephone, fax, ISDN, and other services. It operates in the 100 kbps range based over copper twisted pair wires.
- Higher speed lines used for PBX connections and high speed data services (2 .. 45 Mbps), requiring higher grade cabling.
- Cable TV networks, which can provide higher bandwidth but are designed as distribution networks. At present, CATV networks are unsuitable for interactive applications; technologies to utilize this existing distribution technology for interactive traffic are discussed below.
- Satellite distribution networks, which again are not designed for interactive point-to-point communication and which carry a high end-to-end latency.

These infrastructures support all low-speed data applications as well as TV distribution. Even interactive TV applications can be supported, using cable or satellite networks for the TV distribution part and existing data networks for the control communication. The next evolutionary step is integrated multimedia networks based on the use of ATM cells over cable TV copper networks.

The networking technologies needed to deliver the required throughput over the existing cable infrastructure are already becoming available. One option are hybrid fiber coax (HFC) networks using ATM over fiber backbones and copper cable networks to the end user with a reverse channel of 1 or 2 Mbps. Figure 4–9 shows the architecture of an ATM HFC network [21].

In terms of protocols over an ATM HFC infrastructure, two directions are currently being pursued. One, focusing on the support of data applications over

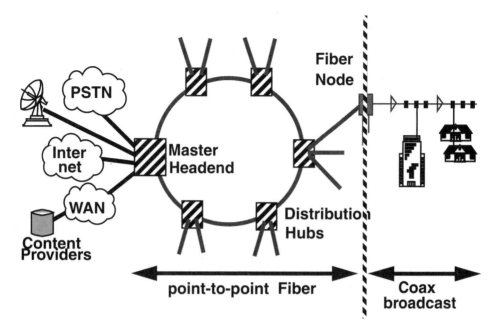

Figure 4–9 ATM HFC Residential Network

CATV networks, is based on the different IP over ATM architectures described later in this chapter. The second, aimed at interactive video distribution, is defining an MPEG-2 over ATM architecture [6][15].

However, bandwidth of several Mbit/s—necessary to support a video quality comparable to existing TV programs and costing a few dollars per hour—lies several years in the future. Nevertheless, ATM over cable TV networks seems to be a promising option for consumer-oriented interactive services of the future.

Another intermediate solution for bringing the throughput required for interactive multimedia to the consumers is the use of High Bit-Rate Digital Subscriber Lines (HDSL) or Asymmetric Digital Subscriber Line (ADSL) technologies delivering synchronous T1/E1 channels over standard telephone cabling [21]. Whereas HDSL provides duplex E1/T1 channels, ADSL multiplexes an E1/T1 downstream channel with a low bit rate (9.6 K ... 16 kbps) duplex control channel on the telephone wires. Since ADSL/HDSL define only the transmission signals, they are not restricted to any given higher layer protocol. ATM might still be the switching technology used above ADSL/HDSL, but ATM overhead is very expensive at these relatively low bandwidths.

In the long run, these technologies seem to be most relevant to those areas where high-grade telephone cabling is available but coaxial CATV cabling or fiber is not.

With a much more immediate perspective than high-quality distributed multimedia, more and more organizations build corporate networks, focusing on voice/data integration. Thus, voice/data integration defines the key requirement for the next generation WANs. Assuming a high-volume bandwidth demand

between the nodes of a corporate network, provision of a high throughput transmission paths is an expensive but feasible task. Given the high transmission
cost, corporate network providers are looking for the most flexible multiplexing
technique to utilize their expensive bandwidth resource as efficiently as possible.
ATM is the most flexible multiplexing technique available, so fiber-based ATM
networks are the logical choice for corporate backbone networks, and most opera-
tors/carriers follow an ATM WAN strategy.

For investment protection, these ATM corporate backbones will have to
interface to almost any existing piece of networking equipment. Further, users
may want to use their corporate network for backbone traffic between major
hubs but still use some other carrier's infrastructure for access to their backbone.
Figure 4–10 outlines the structure of an ATM-based corporate network.

The appeal of this approach is that not only does it reduce transmission cost
today, it also provides a good starting point for a mainly ATM-based communication infrastructure within a corporation.

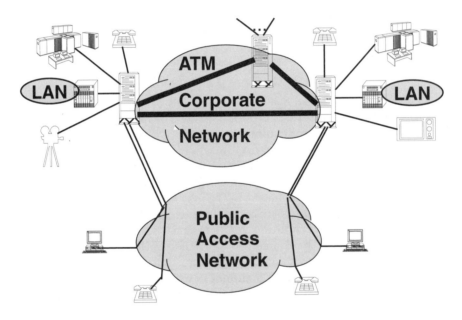

Figure 4–10 ATM-Based Corporate Network

4.3.3   LAN Topology Evolution

To understand the direction, toward which local-area networks are moving going
in the future, it is instructive to look at the evolution of LANs over the last years.
We can identify the following steps.

Linear Topology LANs: The first LANs were single-segment LANs in linear
topology. Ethernets and Token Rings used cables connecting each station to the

next. Moving a system required rewiring, and network operation had to be interrupted to allow the new connection to the network.

Star-Wired LANs: To avoid costly rewiring, wiring concentrators, or multistation access units (MAUs), were introduced. These boxes, now called hubs, were placed in a central location to be efficiently operated by a support team. Once a suitable wiring infrastructure was available, stations could be added by simply plugging another cable in to the hub.

Multisegment LANs: When the number of LANs outgrew the capacity of single-segment LANs, bridges or routers were introduced to interconnect several LAN segments. Since the operation of bridges and routers is crucial to the availability of the total network, they were often installed in a protected area together with the wiring concentrator. With the addition of bridging, routing, and network management functions to LAN hubs, a new generation of boxes, called intelligent hubs, emerged.

Backbone-Connected LANs: With the number of LAN segments and interconnecting bridges growing, topologies became very complex, and control traffic could get very high. One way of structuring the topology is to build a hierarchical network structure based on a backbone network. Shared resources like servers, wide-area connections, etc., are located on the backbone, whereas workgroup communication is mostly kept local on one attached LAN segment. Again, whenever feasible, central resources can be placed with wiring concentrators for easy and controlled access. Eventually, this trend transforms the backbone into the central switching facility of the network.

Switched Backbone or LAN Switching: The next evolutionary step is to replace the backbone by a high throughput switch. Currently, available LAN switches are an evolution from multiport LAN bridges and hence share much of the functionality and design of multiport bridges at higher speeds. An Ethernet switch can be considered as a learning, transparent bridge. To increase efficiency, LAN switches support address screening and filtering. Every port maintains address tables to store the MAC addresses of the attached LAN stations (filtering table) and the MAC address—port number combinations of all known destination stations (forwarding table). During operation, each port learns more and more of the source and destination information and is hence able to route packets to the correct destination port rather than broadcasting them across all attached segments. A typical LAN switch configuration is shown in Figure 4–11.

In this example, some of the ports are dedicated (A..D) to a single station, typically a server, and some ports connect a full LAN segment (E,F,G). With regard to delay characteristics, the LAN switch itself generally forwards frames at media speed. Switches operating in "store and forward" mode add the delay listed in Table 4–6 for frame assembly, assuming a round-robin bus service strategy and a typical 8-port switch.

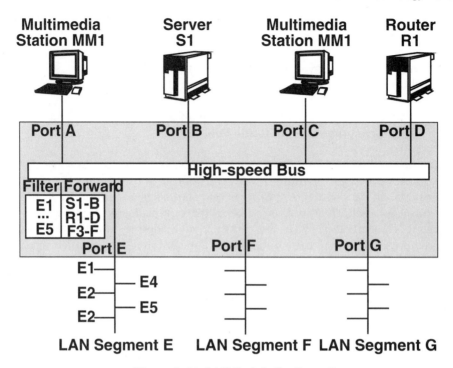

Figure 4–11 LAN Switch Configuration

Worst cases occur when all segments or dedicated stations try to send a maximum size packet at the same time. Given that the switch or bus capacity exceeds the cumulative bandwidth of all ports, all pending packets are forwarded within one packet transmission time; hence, total delay is always between 1 and 2 packet transmission times.

LAN type	max. packet size [byte]	number of ports	delay [ms]
10 Mbit/s Ethernet	1518	8	1 … 2.5
100 Mbit/s Ethernet	1518	8	< 1
16 Mbit/s Token Ring	4096	8	2 … 5

Table 4–6 Store-and-Forward Delay in LAN Switches

However, many switches operate in "cut through" mode, which means frames can be forwarded as they arrive, without being assembled completely on the incoming port. This operating mode further reduces the transmission delay [25]. *Cut-through switching* carries the risk that bad frames can propagate through the switch; however, given the level of reliability today's LANs have, this is generally not a problem. Some switches can adapt from *cut-through* to *store-and-forward* mode, and vice versa, dynamically based on the observed level of faulty frames.

A further performance enhancement comes from the fact that stations on a dedicated LAN segment do not have to compete for access. Hence, a dedicated LAN segment can be operated in full duplex mode, without the need for collision resolution (Ethernet) or token passing on a Token Ring. This mode effectively doubles the throughput to 20 Mbit/s for Ethernet and 32 Mbit/s for Token Ring.

In conclusion, LAN switches can generally improve the performance of LANs, so that the throughput requirements of multimedia applications are satisfied in many practical configurations. However, unless stations are connected as dedicated segments, the same restrictions apply as in the standard LAN environment. LAN switches provide high bandwidth but do not manage bandwidth. Today's products are mostly bus-based architectures with an attractive price/performance ratio, designed for a relatively small number of ports. It is not clear how scalable these designs are. Hence, they are a way to boost the performance of existing LANs rather than a strategic technology for multimedia communication.

4.3.4   Integration of ATM into the LAN Infrastructure

When bandwidth management and other features to provide dependable service guarantees for multimedia traffic are needed, LAN switches are not the solution. In this context, ATM appears to be the promising technology for the near future. For this and other reasons, ATM is generally considered the dominant future networking technology. However, when ATM is introduced into existing LAN infrastructures, the first and most important question is, how can it support existing protocols and applications? As we will not see large "ATM only" networks for a while, we need to look at the question how ATM can interoperate with existing networks. Currently, two different protocol architectures exist to integrate ATM with existing protocols.

4.3.5   Classical IP over ATM (RFC 1577)

The objective of RFC 1577 [22] is to use ATM as one out of many possible IP subnets. In order to do this, one must resolve the following problems:

- Mapping of ATM and IP addresses
- Implementation of the IP address resolution mechanism (ARP) over ATM
- Establishment and termination of direct ATM connections (VCs) between two IP end-systems
- Multiplexing of different protocols over the same ATM connection (VC)

IP/ATM Address Mapping

IP is based on its own hierarchical address structure. ATM uses three variations of the ITU E.164 address format, all totally unrelated to IP addresses. To make matters worse, once a VC has been established, only the VPI/VCI pair (Virtual Path Identifier/Virtual Channel Identifier) is used to route ATM cells.]

IP Address Resolution over ATM

The normal IP address resolution protocol uses the LAN broadcast mechanism to query all listening stations or routers for the MAC address of a given IP address. Since broadcasting is not an appropriate technique in large switched networks, alternative mechanisms are defined in RFC 1577. For ATM SVCs, the ARP-server scheme is used. It requires the existence of a central address server (ARP server) at a known ATM address. At initialization time, every station goes through the following procedure:

1. The IP station builds a connection (SVC) to the ARP-Server.

2. The ARP server then sends an Inverse-ARP-Request to the IP client, which provides its address through a respective Inverse-ARP-Response message. With this protocol the ARP-server builds its address map, relating client IP, ATM, and VPI/VCI addresses.

3. Then, any registered client can query the ARP server for arbitrary IP addresses. If the respective IP station has already been registered, the server will provide the requested ATM address to the client.

4. Knowing the ATM address for a given IP address, the client can then build a direct ATM connection to the target IP station. Further, the client caches the mappings of currently in use or most frequently used address mappings, to off-load the ARP server, which may become a bottleneck, depending on the size of the network.

Multiprotocol Encapsulation

To enable the transporting of several different protocols across the same ATM VC, it is necessary to provide protocol identification information to the receiver. This function is implemented by encapsulating the IP packets into an LLC/SNAP frame, as defined by RFC 1483 [18]. The SNAP, which stands for subnetwork access protocol, provides the required information fields for protocol multiplexing. Thus, IP packets are framed into LLC/SNAP which are then encapsulated into AAL 5 CPCS (Common Part of the Convergence Sublayer) protocol data units.

Note that RFC 1483 defines the use of AAL 5 and not AAL 3/4. RFC 1577 leaves the option either to use null encapsulation over AAL 5, e.g., to frame the IP packets directly in CPSC PDUs, or to use LLC/SNAP encapsulation. If null encapsulation is used, then the used protocol (IP) must be identified via the signalling protocol by proper use of the Higher Layer Compatibility information element.

"IP over ATM"- Connection Management

RFC 1577 does not define the principles of connection management; these are considered local matter. It is up to the end-system to decide what kind of connection to build (traffic descriptor, QoS class) or when to tear it down. This is a policy decision which may be influenced by tariffs, load, or other considerations. However, with the ABR traffic class becoming available, it will be the logical choice for IP traffic over ATM.

In any case, ATM connection management is transparent above the IP layer. Hence, applications have no control over the traffic descriptors or quality of service parameters of the underlying ATM VCs. Further, all traffic streams between two ATM end-systems are interleaved and indistinguishable at the ATM layer. Therefore, RFC 1577 is only suitable for multimedia when bandwidth is abundant and traffic management and control are not needed. This problem will be solved by the flow scheme introduced by the next generation IP protocol (IPv6), which is described in detail in Chapter 5. An extension to the basic RFC 1577 scheme that allows per flow on demand QoS has been proposed in [10].

Further, RFC 1577 handles only the resolution of IP unicast addresses. To support IP multicasting, an extension of the ARP server concept (MARS) has been defined, as described in the next section.

4.3.6   IP Multicasting over ATM

RFC 1577 supports only the resolution of IP unicast addresses. So, despite the fact that the ATM network itself supports point-to-multipoint connections, IP multicasting is not available with RFC 1577. There are two different approaches to resolve this deficiency [9].

The first approach adds a multicast server function to an IP multicast router. In that case, all multicast packets are sent from the sender to the multicast router, which maintains a separate point-to-multipoint VC for every multicast group. The implementation of this concept is very simple. Also, since the MC router is always up-to-date with the state of the MC groups through the use of the Internet Group Management Protocol (IGMP, see section 5.3), maintenance of the multicast VCs is straightforward. The disadvantage is that every local multicast packet has to travel one extra hop, namely, from the sender to the MC server. Furthermore, the MC router may well become a performance bottleneck when a large part of the traffic is multicast traffic. Figure 4–12 shows the flow of packets in this scenario.

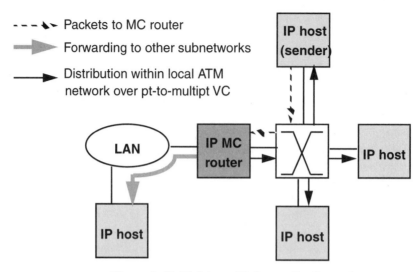

Figure 4–12 Multicast IP Server Configuration

A more general solution is provided by the MARS (Multicast Address Resolution Server) architecture defined in RFC 2022 [2]. The MARS architecture separates the problems of MC address resolution and data forwarding. The MARS itself solves only the problem of mapping or resolving IP group addresses to individual ATM addresses. However, it allows the use of either of two options for the data forwarding part. The first option is the direct VC mesh shown in Figure 4–13

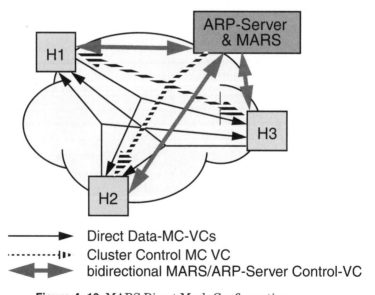

Figure 4–13 MARS Direct Mesh Configuration

The MARS is typically co-located with the ARP-server, which it builds on to handle the IP to ATM address mapping. Although the MARS has complete knowledge about all multicast groups and their respective group members, the management of the multicast VCs is left to the individual clients. In short, a client that wants to send a message to a certain multicast group first queries the MARS for a list of all group members, then builds its own multicast VC, and finally can transmit the data in one hop over its MC VC. This method not only increases the efficiency of the data transfer, it also avoids a server as potential bottleneck. However, since the management of the multicast VCs is now completely distributed, every change in group membership requires several synchronization messages in the network because each group member has to update his tree. The MARS itself uses a cluster control multicast VC to distribute group updates to the members.

The second option is called the Multicast Server (MCS) as shown in Figure 4–14. The objective here is to ease the management of the multicast VCs and at the same time to avoid the potential performance problems of combining the MC server with the MC router.

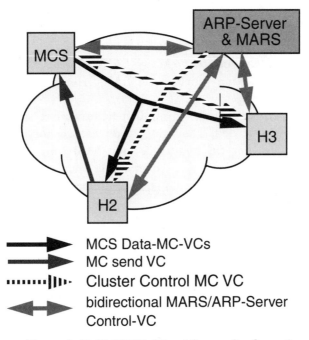

Figure 4–14 MARS Multicast Server Configuration

Again a MARS is used for address resolution; however, data forwarding is provided by one or more multicast servers.

Obviously, this solution introduces additional delay through the additional hop from the sender to the MCS. However, since more than one MCS can be used for different groups, performance concerns are not as bad as in the case of the

MC router. However, in terms of delay, we note that due to the lack of interleaving capabilities in AAL 5, the MCS has to reassemble the packets before it can forward them to the destination. Thus, the MCS adds another packetization delay, which is however not a serious problem in the case of a fast ATM network.

4.3.7   LAN Emulation Services over ATM

Since IP is not the only protocol being used in the local area, a more general solution than RFC 1577 supporting all existing LAN protocols over ATM is desirable. To speed up the deployment of local area ATM, such a function is necessary since the majority of existing LAN applications are based on protocols like IPX or NetBIOS. If use of ATM required the adaptation of existing applications, ATM would be confined to a niche market; thus, a solution that allows all existing applications to run unmodified over ATM is required.

This solution is called "LAN Emulation" (LANE) over ATM [3]. The basic idea is to provide a LAN MAC layer interface on top of ATM AAL 5 services. For LANE, problems similar to those in the *"IP over ATM"* case must be solved:

- *Addressing:* The mapping between MAC addresses and ATM addresses must be defined. Address resolution can be implemented by an address server-based scheme similar to RFC 1577.
- *Multicasting-/Broadcasting:* Many LAN applications utilize the multicasting or broadcasting functions of LANs. Hence, a broadcast server is required.
- *Connection Management:* An implicit (transparent) connection management scheme is needed. The first packet can be forwarded by the LES server to circumvent the relatively long ATM connection establishment delays. All later packets will go over direct connections between the two end-stations. Connection take-down is again local matter.

The implementation of a LES solution distributes the LANE functions among the clients (LECs), the LES server (LES), and a Broadcast and Unknown Server (BUS), which handles all broadcasts traffic in the emulated LAN. The different components are connected by virtual circuits (VCs), as shown in Figure 4–15. LECs, LES, and BUS communicate over switched virtual circuits, using AAL 5 encapsulation. Here, there is no need for protocol multiplexing because, this function is available above the MAC layer.

The LANE operation consists of the following steps:

1. *Initialization.* An LEC establishes the default VCC to the LEC.

2. *Address Registration.* The LE client notifies the LE server of its MAC and ATM addresses.

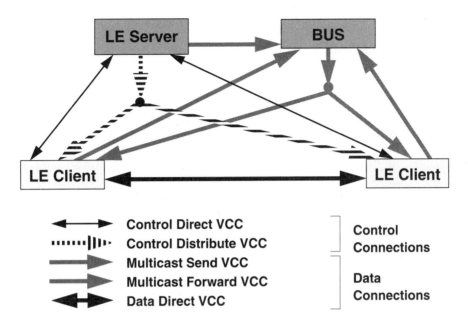

Figure 4–15 LANE Configuration

3. *Address Resolution.* The LEC learns the ATM address of a target MAC address from the LES and establishes direct VCCs to the target MAC address. The first address a client queries from the LES is the address of the Broadcast and Unknown Server (BUS). After the client has established the connection with the BUS, it can broadcast data via the BUS or use the BUS broadcast to forward frames to other clients to which it not yet has established a direct VC.

4. *Data Transfer.* The LEC software encapsulates data with the respective MAC frame, sends the first packet over the default VCC, and sends further packets over the direct VC.

5. *Multicasting-/Broadcasting.* The LEC sends broadcast frames to the BUS, which then forwards them over its multicast forward VC if available or replicates data and transmits over the default VCs to all its clients. Thus, the BUS provides a nonselective broadcast within the emulated LAN, where all clients receive any multicast data and then have to filter out the relevant data based on their higher-layer group address.

6. *Connection Management.* Direct VCCs can be terminated either explicitly by the client or by timer control. Again, below the MAC layer, no information of required QoS or traffic characteristics is available. So, as with RFC 1577, a LANE client will normally use the ABR traffic class if available.

Similarly to RFC 1577, no QoS support is built into LANE. In fact, LANE shields any QoS information available at higher layers through the use of the LAN MAC interface from the lower ATM layer. Hence, LANE is, like RFC 1577, not an architecture to support multimedia communication.

4.3.8   Native Multimedia Communication over ATM

One of the main motivations for using ATM is the support for different Qualities of Service (QoS) over the same network. This is a feature specially required by multimedia applications. Both architectures, RFC 1577 and LANE, shield the ATM signalling and thus the QoS control from the user. Their purpose is to provide compatibility of existing data protocols with a new network, as well as defining the interworking between existing and new networking technologies.

A third alternative is a direct ATM service interface allowing explicit QoS selection and control by the user. Such a multimedia enhanced or native ATM API is still under discussion. The X/Open group is defining such interfaces based on either XTI or the socket interface. A second approach, already commercially available, is the "Native ATM Socket Interface" in IBM's AIX™ Vs. 4.2 operating system [11].

This interface extends the popular socket API to allow applications to pass a traffic management control structure to the adapter in order to set up SVCs with the required QoS. When using "native ATM sockets," data can be exchanged only between homogeneous, e.g., ATM, end-systems. The approach is general enough to cover networks other than ATM, which may today or in the future provide QoS support. However, there is no interoperability between such networks since no network layer protocol, but only AAL 5, is used to transport the data between the different end-systems.

The great advantage of this interface is that not only does it provide QoS enhanced access to ATM networks, it also allows existing multimedia socket applications to be ported with minor modifications in the connection establishment code to support QoS selection over ATM networks.

4.4 Summary

Current local area networks provide enough throughput to develop experimental distributed multimedia systems. However, even compressed 1.4 Mbps video streams can fill up a 10 or 16 Mbps LAN very quickly. In the wide area, the throughput problems are even worse. However, throughput alone is not enough to satisfy the requirements of packetized multimedia communication. New networking technologies, like ATM, are better suited to satisfy the delay requirements of isochronous streams than are the traditional packet LAN and WAN technologies. In addition, ATM has the advantage of being equally suitable for the local and wide area. Thus, we expect to see more and more switched networks, many of them based on the ATM switching technology. To protect existing cabling and adapter investments, cabling and adapters will not only be based on high-speed (100 or 155 Mbps) fiber links, they will also use existing LAN technology as transmission media. Thus, although the user network interface may still look like an Ethernet or a Token Ring, bandwidth will be dedicated and sufficient for most applications. With no access competition, access and transmission delays can be minimized so that multimedia applications can be supported appropriately.

References

[1] A. Albanese and M. Luby, "PET – Priority Encoding Transmission," In *High-Speed Networking for Multimedia Applications*, edited by W. Effelsberg, O. Spaniol, A. Danthine, D. Ferrari. Dordrecht: Kluwer Academic Publishers, 1996, pp.247–265

[2] G. Armitage, *Support for Multicast over UNI 3.1 Based ATM Networks,* IETF RFC 2022, November 1996.

[3] ATM Forum, *LAN Emulation* 1.0, Edited by B.Ellington, 3/1995.

[4] ATM Forum, *ATM User-Network Interface Vs.3.1,* Prentice Hall 1995.

[5] ATM Forum, *ATM User-Network Interface Signalling Specification Vs.4.0.*

[6] ATM Forum, Technical Committee Audiovisual Multimedia Services (AMS): *Video on Demand Specification 1.*0, af-saa-0049.000, 12/1995.

[7] E. W. Biersack, "Performance Evaluation of Forward Error Correction in ATM Network," *Proceedings of the ACM SIGCOMM 1992, Computer Communication Review* 22(4):248–257, 1992.

[8] C. Bisdikian, B. Patel, F. Schaffa, and M. Willebeck-LeMair, "The Use of Priorities on Token-Ring Networks for Multimedia Traffic," *IEEE Networks* 9(5):28–37, Nov./Dec.1995.

[9] T. Braun, S. Gumbrich, and H. Stüttgen, "Comparison of Concepts for IP Multicast over ATM," *Proceedings of the 2nd IEEE ATM Workshop*, San Francisco, August 1996.

[10] T. Braun, H. J. Stüttgen, "Implementation of an Internet Conferencing System over ATM," *Proceedings of the 3rd IEEE ATM Workshop*, Lisbon, May 1997, pp.287-294

[11] D. Chang, W. Hymas, S. Sharma, H. Stüttgen, and S. Wise, "Socket Extensions for Native ATM Access," *Proceedings of the 2nd IEEE ATM Workshop*, San Francisco, August 1996.

[12] S. E. Deering, *Host Extensions for IP Multicast*, IETF RFC 1112, 1989.

[13] L. Delgrossi, D. Hehmann, R. G. Herrtwich, C. Halstrick, O. Krone, J. Sandvoss, and C. Vogt, "Media Scaling for Audiovisual Communication with the Heidelberg Transport System," *Proceedings of the 1st ACM Multimedia Conference*, Anaheim 1993.

[14] M. dePrycker, R. Peschi, and T. Van Landegem, "B-ISDN and the OSI Protocol Reference Model," *IEEE Network* 7(2):10–18, March 1993.

[15] S. Dixit and P. Skelly, "MPEG-2 over ATM for Video Dial Tone Networks: Issues and Strategies," *IEEE Network* 9(5):30–40, September 1995.

[16] P.W. Edwards, R. E. Hofman, T. Liaw, et al., I *Ipsilon Flow Management Protocol Specification for IPv4 Vs. 1.0*, IETF RFC 1953, May 1996.

[17] D. Hehmann, M. Salmony, H. J. Stüttgen, "Transport Services for Multimedia Communications on Broadband Networks," *Computer Communications* 13(4):197–203, May 1990.

[18] J. Heinanen, *Multiprotocol Encapsulation over ATM AAL 5*, IETF RFC 1483, July 1993.

[19] *IEEE Network* 6(5), September 1992, Special Issue on *Congestion Control in ATM Networks*.

[20] Ipsilon: IP Switching: The Intelligence of Routing, the Performance of Switching. URL: http://www.ipsilon.com, February 1996.

[21] T. Kwok, "A Vision for Residential Broadband Services: ATM-to-the-Home," *IEEE Network* 9(5):14–28, Sept. 1995.

[22] M. Laubach, *Classical IP and ARP over ATM*, IETF RFC 1577, January 1994.

[23] D. LeGall, "MPEG: A Video Compression Standard for Multimedia Applications," *Communications of the ACM* 34(4):47–58, April 1991.

[24] M. Liou, "Overview of the p*64 kbit/s Video Coding Standard," *Communications of the ACM* 34(4):60–63, April 1991.

[25] R. Mandeville, "Ethernet Switches Evaluated," *Data Communications* pp.66–78, March 1994.

[26] R. Nagarajan and C. Vogt, *Guaranteed Performance Transport of Multimedia Traffic over the Token Ring*, IBM ENC TR 43.9201, 1992.

[27] D. Newmann and B. Levy, "The Real Fast Ethernet," *Data Communications* pp.67–78, March 1996.

[28] P. W. Edwards, R. Hinden, P. Newman et al., *Ipsilon's General Switch Management Protocol Specification Vs. 1.0*, IETF RFC 1987, Aug. 1996.

[29] R. M. Newman, Z. L. Budrikis, and J. L. Hullet, "The DQDB MAC," *IEEE Communication Magazine* 26(4):20–28, April 1988.

[30] Y. Rekhter, et al., Tag Switching Architecture. *URL:* http://www.cisco.com, April 1997.

[31] F. Ross, "An Overview of FDDI" *IEEE Journal of Selected Areas in Communication* 7(7):1043–1051, Sept. 1989.

[32] G. Watson, A. Albrecht, J. Curcio, D. Dovem, S. Goody, J Grinham, M. P. Spratt, and P. A. Thaler, "The Demand Priority MAC Protocol," *IEEE Networks* 9(1):28–34, Jan. 1995.

[33] D. J. Wright, *Broadband: Business Services, Technologies, and Strategic Impact,* ISBN 0-89006-589-6, Artech House Inc. 1993.

Network and Transport Layer Protocols for Multimedia

Wolfgang Effelsberg

*I*n this chapter, we discuss network and transport layer issues for multimedia communication. In the layered architecture of communication systems, still very widely accepted today, the tasks of the network and transport layers are well defined: the network layer is responsible for setting up *routes* from a source node to a destination node, and the transport layer handles end-to-end issues between processes running on these nodes, such as *reliability*. Together the network and transport layers establish a "data pipe" between a process on the source computer and a process on the destination computer. Obviously, these functions are still necessary in multimedia communications, but we will have to take the new communication requirements introduced in Chapter 2 into account.

We briefly describe how traditional network and transport layer protocols work. We show why they are inappropriate for continuous data streams. We present new algorithms and protocols to support continuous streams, and we give examples of new protocols where these new principles are implemented.

5.1 Principles and Algorithms of Traditional Protocols

Computer networks were initially designed to carry non-real-time traffic, in particular for remote login, electronic mail, and file transfer. Many design decisions in traditional protocol stacks are based on the assumption that bounds on delay and jitter are not an issue. As a consequence, traditional networks, such as the current Internet, are not able to handle multimedia traffic well. We now investigate what the major functions and algorithms in the traditional network and transport layers are, and why they are inappropriate for real-time data streams.

5.1.1 Routing

Routing is the main task of the network layer. Only in the most trivial local area networks where every node is physically connected to every other node is routing unnecessary; in reality, all major networks consist of an interconnection of subnetworks. We can depict the network as a graph of nodes, representing subnetworks, and edges, representing the links between them. The problem is now to find an *optimal path* from a given source node to a given destination node.

Routing actually involves two major subproblems: one is to find an optimal path in the routing graph, under changing network loads and perhaps even a changing network topology, and the other is to get all incoming packets through a router at runtime in an optimal way. In order to find an optimal path, all nodes of a routing domain have to cooperate, using the same optimization criteria; otherwise, packets might never reach their destination. In contrast, packet scheduling within a node can be handled differently in each node. Their internal architectures, the operating system software, the priority schemes, etc., can all differ.

There are two fundamental approaches to routing: either the pathfinding algorithm is executed every time a packet is injected into the network, or a path from source to destination is computed only once for the duration of a connection. In the first approach, each packet finds its way independent of other packets, similar to a telegram in the postal service: thus, this is called the *datagram* technique. In the second approach, all packets of a connection follow the same path through the graph; the connection-setup packet leaves a trace with routing information in each node on the path, basically consisting of a path identifier and an output port, and all subsequent packets follow the same path. This technique is similar to a telephone call where a fixed route through telephone switches is used for the duration of a call; thus, this is called the *virtual circuit* technique. The main advantages of datagrams are:

- Efficiency for short connections (no connect and disconnect phases in the protocol)
- Robustness (no state information in the nodes, no cleanup in the event of failure)
- Easy internetworking (very little "consensus" between participating subnetworks required)

The main advantages of virtual circuits are:

- Efficient routing at runtime (no pathfinding algorithm to be executed during a connection)

- The ability to use call acceptance control to avoid network congestion (a new call is rejected when the network is overloaded)

In the early days of networking there were vivid international debates as to whether datagram or virtual circuit technology would be better at the network layer. The Internet architects in the United States U.S. voted for datagrams (connectionless operation), whereas the international standards for packet switching networks ended up with virtual circuits (connection-oriented operation). In particular, CCITT Recommedation X.25 is based on virtual circuits, and so is Frame Relay. Thus, the packet-switching services offered by telecom carriers to their customers have always been connection oriented.

In the ISO/OSI protocol stack, the network layer is typically connection oriented, namely, based on X.25, so the most widely used version of the ISO transport layer, called Class 0, is almost empty. The only major function in ISO TP_0 is packet segmentation and reassembly.

For compatibility with the TCP/IP approach, ISO later added a connectionless version of the network layer protocol, and there is a corresponding version of the ISO transport protocol, called Class 4, with all the reliability functions built in. The algorithms used in TP4 are again very similar to those in TCP, i.e., error detection and retransmission for bit error recovery, sequence numbers for packet loss recovery, and sliding window for flow control. The combination of ISO connectionless network layer and TP4 is almost identical to TCP/IP.

It is important to understand that the service offered to the user at the upper interface of TCP or ISO TP0 or ISO TP4 is practically the same, namely, reliable transmission of a byte stream from process to process. Therefore, it is easy to run high-layer application protocols over either TCP or ISO TP0 or TP4.

All traditional network and transport layers support peer-to-peer communication only, i.e., a connection is always set up between exactly one sender and one receiver, and a packet always travels from one sender to one receiver. So all routing algorithms are 1:1. The principle of a traditional packet router is shown in Figure 5–1.

Let us now take a closer look at the routing algorithms (pathfinding algorithms) in use today. They can be classified into *static routing* and *adaptive routing*. In static routing, all routes are precomputed for a given topology and are independent of the current network load. In adaptive routing, the pathfinding algorithm automatically takes into account new or obsolete nodes and links into account as well as the current load of nodes and links. In adaptive routing, it is not easy to find a global optimum, nor is it easy to find robust distributed algorithms and protocols to implement the routing table updates.

In static routing, each node has a table with entries in the form (source, destination, outgoing link). An incoming packet contains the destination address (or a connection identifier, in the case of virtual circuits), and the routing decision is reduced to a quick table lookup. When the network topology changes, a network control center recomputes the global routing table, and the new table is downloaded into all nodes. Well-known algorithms, such as Dijkstra's Shortest Path algorithm [17], can be used to compute a global optimum when the network

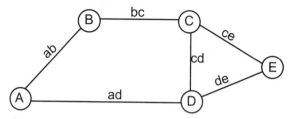

(a) Network topology for our example

RT	Routing table at C		

From C to	link	cost
A	bc	2
B	bc	1
D	cd	1
E	ce	1

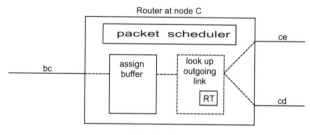

RT = Routing Table

(b) Traditional packet router

Figure 5–1 A Traditional Packet Router

topology is represented as a partially connected graph with weighted edges. The advantages of static routing are simplicity and speed at runtime; its disadvantages are inflexibility, far-from-optimal decisions in the case of topology updates or dynamic load changes, and the performance and reliability problems inherent to all centralized algorithms. Many X.25 networks worldwide use static routing, as does SNA, IBM's Systems Network Architecture.

In adaptive routing there is no central node; each node gets some limited information from neighboring nodes and/or extracts information from packets underway. A simple example is the Backward Learning algorithm, illustrated in Figure 5–2. For routing purposes, each packet contains a hop counter in its header, counting the number of links traversed so far. If a packet arrives at node E from source node A over link ce and has traveled 3 hops, there must be a path from E back to A via ce with a distance of 3. Perhaps a little later, a packet arrives at E via de with a hop counter of 2. The new path ad-de is shorter, in terms of hops, than the old path ab-bc-ce, and E updates its routing table.

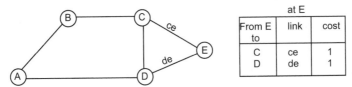

(a) E only knows its immediate neighbors

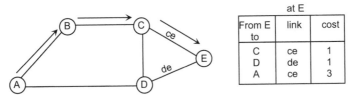

(b) First packet has arrived from A on link ce

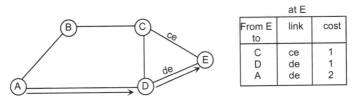

(c) A new packet has arrived from A on link de

Figure 5–2 Backward Learning

Unfortunately, very simple adaptive algorithms such as Backward Learning seldom work well. For example, how do we find routes to nodes from which we have never seen a packet? Or, if a node is removed from the network, this information will obviously never be propagated to other nodes. A solution to the latter problem is to attach a time-out to each routing table entry. But what would be the right time value? The adaptive routing algorithms in use today are more complicated and more powerful.

Finding Shortest Paths

A fundamental problem common to all routing algorithms is to find the shortest paths between all pairs of nodes for a given topology, in other words, to find all shortest paths for a given graph with weighted edges. Fortunately, a number of efficient algorithms for this problem exist. The earliest one was published by Dijkstra in 1959. It finds all shortest paths for a given start node. It is still very widely used today, and so we explain it briefly.

Our graph consists of a set of nodes, N, and a set of edges, E. We proceed as follows:

1. Initialize the set of known paths P with N_0, the start node.

2. Consider all immediate neighbors of nodes that are already part of the set of known paths P; each path ending in such a node is a candidate path.

3. Sort the candidate paths by length (i.e., sum of link weights).

4. Add the shortest candidate path to P.

5. If there are unconnected nodes left, continue with step 2.

When the algorithm terminates, P is the set of shortest paths to all nodes starting at node N_0.

Dijkstra's algorithm is often called "Shortest Path First" (SPF). So, if we know the full topology of our network and have a good metric assigning weights to our links, we can apply the SPF algorithm to find the optimal routes. This is typically done in static routing, in particular in X.25 and SNA networks.

In adaptive routing, the SPF algorithm is not as easy to use because SPF requires knowledge of the full network topology. One possibility is to have a central node, the Routing Control Center (RCC); all nodes report the status of the links in regular intervals to the RCC. The RCC computes the new topology and broadcasts it to all network nodes. But a central resource is a bad idea in almost all distributed systems; it can easily become a performance bottleneck, and it is a single point of failure. The alternative is to design truly distributed topology update algorithms: a topology "database" is maintained in each node and updated by incoming link state messages. In fact, the OSPF (Open Shortest Path First) algorithm of the Internet falls into this category.

Distance Vector Routing

The first routing algorithm that was widely deployed in the Internet is called *Distance Vector Routing*. Each node maintains knowledge of its shortest distances to other nodes of the network. A table contains an entry for each known node, the distance to that node, and the link to be used to get there (in fact, this is the routing table already presented in Figure 5–1). The table is therefore also called the Distance Vector.

For simplicity, let us assume that distances are counted in hops, i.e., the distance is always 1 for immediate topological neighbors. If a node receives a distance vector update message from a neighbor, it will update its local routing table as follows:

- If a new node name appears, then add the entry to the local table, with a distance of i+1 (one hop more to reach the neighbor).
- If the message shows that there is a shorter path to a known node through the neighbor, then update the local table entry accordingly.
- If an arriving entry leads to a route longer than one that is already known, ignore it.

- If an arriving entry has distance = ∞, then drop the local entry for that destination node (a link or node has become unavailable on the path via that neighbor).

The operation of distance vector routing for the topology of our example is illustrated in Figure 5–3. Let us assume that only nodes A, B, C, and D formed the network so far. When node E joins the network, the immediate neighbors C and D establish local routes to E and update their local routing tables accordingly. Nodes A and B don't yet know E. After one exchange of routing table update messages, they will have learned from C and D, respectively, about the new node E.

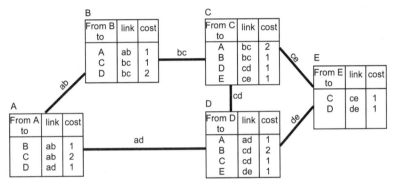

(a) E has just joined the network

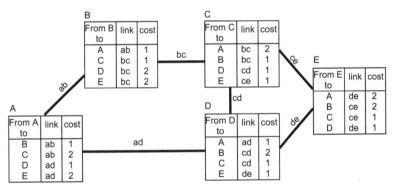

(b) After one round of routing table update messages

Figure 5–3 Distance Vector Routing Algorithm

It is straightforward to extend the distance vector routing algorithm to include weighted links. Instead of setting the cost of every link to 1, we can use different weights for the links, reflecting different delays or bandwidths or even current utilization. If we assigned a weight of 1 to ab, a weight of 2 to bc, a weight of 1 to cd, and a weight of 6 to ad in our example, the routing tables at A and D would route packets via B and C rather than use the direct link ad. It is also possible to use alternate routes in parallel, chosen with probabilities inversely proportional to their weights.

   Distance vector routing is very easy to implement, very efficient, and still very widely used in the Internet today. However. it has two major drawbacks:

- Bad news propagates slowly.
- Routing loops can occur in transition periods.

Let us assume that links bc and ad go down in our example. Nodes A, B, C, and D will discover the failure immediately and will set their direct distances to ∞. However, the entry (C, ab, 2) at node A remains unchanged. In the next round of routing table updates, let us assume that B receives the current distance vector from A first, indicating a distance of 2 from A to C. Since its own distance to C is currently ∞, it will update its entry to (C, ab, 3). The fact that bc is down is no longer propagated. From now on, packets from B to C will bounce back and forth between B and A. In the next round of updates, A will set its entry to (C, ab, 4), and so on. Thus, the fact that bc is down leads to a stepwise incrementation of the distances at A and B until a predefined maximum (say, the diameter of the network) is reached. Only then is the distance (A, C) set to ∞ at nodes A and B. This procedure is called "counting to infinity."

   Obviously, for a large network, it can take many rounds until all nodes understand the new topology. And in the meantime, routing tables can be inconsistent, causing loops and "bouncing packets." Looping data packets (IP packets in the Internet) will be thrown away once they have reached their "time to live," so there is at least no infinite looping, but considerable additional network load and delay are created. An interesting remedy was discovered in 1989 [10].

   For actual routing in the Internet, a protocol for exchanging the distance vectors among nodes must be defined. The Internet family of protocols for this purpose is called RIP, the Routing Information Protocol. Typically, RIP protocols are quite simple and very reliable. For an excellent discussion the reader is referred to [28].

Full-Topology Routing

The fundamental problem with distance vector routing is that each node has only partial information about the network: in the steady state, it only has some knowledge about distances but does not understand the full topology of the network. Of course, if each node knew the full topology, it could always recompute all optimal routes (i.e. shortest paths) locally, in particular once it has heard of a link or node failure.

   This is the basic idea of another class of algorithms which we call "full-topology routing." At each node, the shortest paths to all other nodes are computed, using Dijkstra's SPF algorithm; the Internet community often calls it the "Open Shortest Path First" algorithm (OSPF). Another name for the same family of algorithms is "Link State Algorithms" because all nodes have full knowledge of all link states.

In full-topology routing, each node maintains a topology database, reflecting the state of all nodes and links. The routing table update messages now report changes in that topology; this scheme is obviously less costly than retransmitting the full topology in every round. The advantage is that all nodes will better understand *where* the problem is and can recompute *all* their shortest paths accordingly. Node A in our example will find out that *bc* is down after receiving the first topology update message from *B*. It will then no longer try to route packets to *C* via *ab*. No packets will bounce between *B* and *A*. Bad news will never take more than *d*-1 rounds to reach everyone if *d* is the diameter of the network. The operation of full-topology routing for our example is illustrated in Figure 5–4.

A disadvantage of full-topology routing is that more information has to be maintained locally, requiring more memory in the router, and that recomputing all shortest paths is more CPU-intensive than updating routing tables based on

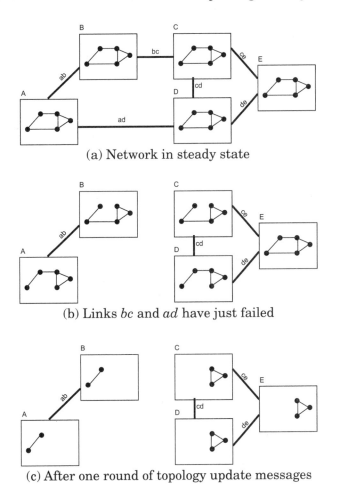

(a) Network in steady state

(b) Links *bc* and *ad* have just failed

(c) After one round of topology update messages

Figure 5–4 Full-Topology Routing Algorithm

distance vectors. But modern host and router hardware can easily accommodate these requirements. Another disadvantage is that the distributed topology update algorithm is much more complicated than just sending distance vectors: there are many different message types to be exchanged between nodes, making implementations more error-prone. In any case, the OSPF class of algorithms is considered by many to be superior to distance vector routing and is now generally recommended by the Internet architects.

Research into routing algorithms for the Internet is still a very active field; for example, a very efficient algorithm combining ideas of distance vector routing and link state routing is DUAL ("Diffusing Update Algorithm") [25]; variations of this algorithm are now implemented in Internet router products. An ingenious method to prevent loops in transition phases was proposed in [26]. Many other ideas for improved routing can be found in the literature.

The actual routing algorithms and protocols used in the Internet have gone through a series of revisions and are quite sophisticated. In fact, Internet routing distinguishes between hosts and subnetworks; a subnetwork is often under the control of one administration (e.g., a university or an enterprise). Internal routing is done within a subnetwork, and external routing is done between subnetworks. The advantage is that routing tables no longer have to understand all routes to all hosts but only those to foreign subnetworks. This keeps them smaller and makes updates much less frequent. Different protocols and control messages are used for internal and external routing. The details are beyond the scope of this book; the reader is again referred to [28].

5.1.2 Reliability

By reliable data transfer we mean the delivery of all packets without errors and in sequence. In traditional network and transport layer protocols, the main reliability functions are implemented as follows:

- Bit error recovery is based on *error detection and retransmission*. The sender adds just enough redundancy to detect bit errors, but not enough to correct them at the receiver. If a bit is spoiled during transmission, the receiver detects the error and the packet is retransmitted. The sender learns about the bit error in one of two ways: Either the receiver sends a negative acknowledgment (NACK), or the sender signals a time-out unless a positive acknowledgement (ACK) is received within a predefined interval. TCP and ISO TP_0 actually use the latter variant, called "Positive Acknowledgment or Retransmission."

- *Packet loss recovery* is based on sequence numbers. The sender numbers all packets consecutively, and a missing packet is detected in the same way as an erroneous packet, either by the receiver's NACK or by the sender's time-out.

- *Flow control* is done by the Sliding Window mechanism. The purpose of flow control is to prevent slow receivers from being overrun by fast senders. The sliding-window algorithm for flow control allows the sender to transmit packets at its own speed until a window of size w is used up [47]. It then has to stop and wait until acknowledgments from the receiver open the window again. In the best case, the acknowledgments will have arrived before the window size has been reached, keeping the flow of packets steady. However, if the receiver is too slow or network congestion has delayed the delivery of packets, the sender will have w packets in transfer and will stop sending until the next ACK has arrived. Actually, in the TCP protocol, w is not counted in terms of packets but in terms of bytes in transfer; however, the principle is the same.

Let us first take a closer look at the principles of bit error recovery. A fundamental (and intuitive) observation is that the error recovery power depends on the amount of redundancy added to the packet. The more bits we add, the higher the chance that we can detect or even correct an arbitrary number of bit errors. Error-detecting and error-correcting codes are very interesting fields of research, but for details we refer the reader to the literature [33]. In any case, we need many fewer bits to *detect* errors (we know *that* an error has occurred) than we do to *correct* errors (we know *where* an error has occurred).

When TCP/IP and the ISO protocols were designed, communication lines were typically telephone lines, with low bandwidth and high bit error rates, and computers were very fast compared to transmission speeds. Given the parameters valid in these times, the information theory experts showed that the use of error-detecting codes and retransmission was better than the use of error-correcting codes. Thus, all the lower layer protocols (also those at layer 2) are based on error detection and retransmission. Today, we use fiber optics whose bandwidth is high and bit error rate is low. The speed of our processors, however, has not increased as quickly as the transmission speed, and, in the next section, we will reconsider this fundamental design decision.

Let us also take a closer look at the Sliding Window flow control protocol. It was invented as a major improvement to a simple stop-and-wait algorithm where the sender sends one packet at a time and waits for an ACK. The main advantage of the sliding window is that bandwidth is not wasted by unnecessary waiting, not even on long distance links; a fast sender will always keep a slower receiver busy. Also, there is flexibility in the frequency in which ACK packets are generated: after every packet received or after every third packet received, opening the window by 3, etc. The separation of mechanism and policy offers the potential for optimization. The Sliding Window protocol is illustrated in Figure 5–5 for a window size w of 3.

5.1.3 Multicast

The term *multicast* defines a stream of data flowing from one sender to many, but not all nodes in the network. Multicast is somewhere in the middle between a *peer-to-peer* connection with exactly one sender and one receiver, and a *broad-*

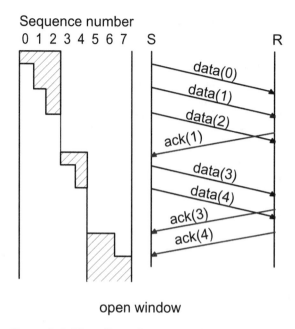

Figure 5–5 Flow Control with a Sliding Window of Size 3

cast connection where all receivers known to the network get a copy of each packet.

For many applications, a multicast mechanism in the network is highly desirable. Let us use e-mail as an example. If an Internet user in Los Angeles wants to send electronic mail to four colleagues, one in Chicago, one in Boston, one in New York, and one in Germany, he or she lists all four of them as recipients of a single mail. The local mail tool then creates four copies of the mail, sets up four different TCP/IP connections, and transmits the four mails separately. This implies that all four copies travel over the links separately. Bandwidth could obviously be saved if intermediate mailers on the path, e.g., the one in Boston, were able to understand the optimal multicast tree and were able to make the necessary copies as the message travels along. If multicast is available within a network architecture, the unnecessary duplication of messages at the application level, typically at the source node, can be avoided.

Unlike the X.400 mail architecture [40] proposed by ITU-T (the former CCITT), the Internet mail architecture and the SMTP mail protocol lack this capability. So even for very large mailing lists, all necessary copies are created at the originator's site and are transmitted separately. While this might be acceptable for text messages, it is prohibitive for the very high data rates of digital audio and video streams. Figure 5–6 illustrates the issue.

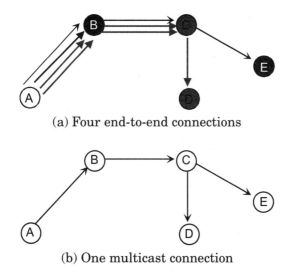

(a) Four end-to-end connections

(b) One multicast connection

Figure 5–6 Motivation for Multicast Support within the Network

Multicast in LANs

In contrast to layers 3 and 4, multicast has always been part of the lower layers in LAN architectures. Multicast addressing is already defined in the IEEE Logical Link Control layer for LANs (IEEE 802.2) where the first bit of an address indicates whether it is a single-station address or a group address (see Figure 5–7). Address recognition and frame copying are implemented on network adapters for CSMA/CD, Token Ring, Token Bus, and FDDI, and each station can be programmed to copy not only frames with its own single-station address, but also those of a given list of group addresses.

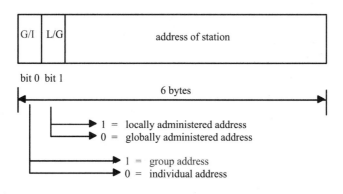

Figure 5–7 LAN Address Format According to IEEE 802

Multicast is easy to implement in single-segment LANs, given their broadcast topologies: On a bus or ring, each frame reaches all stations on the medium automatically. There is no multicast tree topology, and no multicast routing tables are necessary. On broadcast media, the multicast function is thus reduced to group address recognition in each station, and group address management is a network management issue. While group address formats have been part of the LAN standards since the very beginning, group address management was never well supported. Fortunately, an IEEE expert group (802.1Q) is now working on group address management protocols such as GARP and GMRP. They are also addressing networks of bridged LANs.

Multicast in ATM

ATM, the Asynchronous Transfer Mode, is described in Chapter 4. The ATM packets are called *cells* and have a fixed size of 53 bytes. An ATM connection is a virtual circuit similar to an X.25 connection: All ATM switches on the path from the sender to the receiver route cells explicitly, using a switch fabric. When a point-to-point connection is established, a "mini cell route" is established in each ATM switch along the virtual circuit.

Adding multicast to an ATM switch thus requires a cell duplication function within the switch and an architecture to map multicast (i.e., group) addresses to "mini cell routes" in the switches. The cell duplication function is often straightforward: the principle is that the ATM switch can be "programmed" to duplicate cells arriving on a particular virtual circuit and route them to a specified number of outgoing virtual circuits.

Several different architectures for cell duplication in ATM have been proposed, and an overview can be found in [19]. Two simple examples are shown in Figure 5–8.

A multicast connection in an ATM network consists of a tree of ATM switches. The derivation of an optimal multicast tree for a given group address is quite difficult; tree routing requires detailed knowledge of the global link topology, quality-of-service (QoS) parameters of the multicast connection, current load of the links and switches, etc. Whereas most ATM switches available in the market today support cell duplication, the tree routing problem and especially the signaling protocols for multicast are not yet implemented. If multicast is used at all today, the multicast connections are often set up "by hand," and are typically in the form of permanent virtual circuits. Only recently, new protocols for multicast were standardized; UNI 4.0 defines multicast signalling and a leaf-initiated join procedure. A more detailed introduction to multicast in ATM was presented in Chapter 4.

5.2 Problems with Traditional Protocols

As we have seen, the most important new requirements of multimedia communication are:

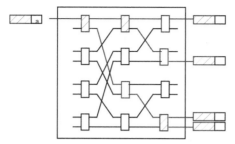

(a) Cell duplication within the switch fabric

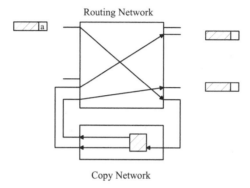

(b) Cell duplication with a copy network and re-routing

Figure 5–8 Multicasting by Cell Duplication in ATM Switches

- High bandwidth
- Support of Quality of Service
- Multicast

A number of application examples with their typical Quality-of-Service requirements can be found in [24].

Chapter 3 introduced the fundamental compression techniques for audio and video, and we have seen that we can transmit high-quality audio and video at amazingly low data rates. Chapter 4 showed that all modern subnetwork technologies provide bandwidths of 100 Mbps and more. So, we can consider the bandwidth problem solved.

But we have also seen that traditional network and transport layer protocols did not support the QoS concept, not in the Internet protocol stack nor in the ISO/OSI stack nor in any of the proprietary protocol stacks of the major vendors. And multicast at the network layer was not supported, either. We now address these two issues in separate subsections.

5.2.1 Traditional Protocols Cannot Guarantee Quality of Service

Three examples from layers 2, 3, and 4 show us how today's algorithms spoil the isochronous flow of packets.

- With the LAN medium access control protocol of CSMA/CD (Ethernet LANs), a station sends a packet immediately when the medium is free. If the station senses another packet transmission in progress, it waits until the other transmission is over. If two stations find the medium free and start to send at the same moment, the packets collide on the medium, and both stations interrupt the transfer and start again after a random time interval. An unbounded variable delay is introduced into the data stream by the MAC algorithm. Similar arguments apply to Token Ring, FDDI, and most other MAC algorithms in LANs; even if the maximum delay is bounded, there can be a high degree of delay jitter.
- Error correction by retransmission is harmful: Obviously, the delay of a retransmitted packet is higher than that of a correct packet, introducing a considerable amount of jitter.
- Flow control by sliding window also introduces considerable jitter: When the sender is faster than the receiver, it sends packets quickly until the window size w is reached; then, it stops and waits until acknowledgments from the receiver arrive and then sends again a batch of packets.

There are many more examples of algorithms in the traditional protocol stacks that are harmful for QoS-controlled networks. We conclude that *we have to design and implement new or modified algorithms for many major protocol functions.* That is the reason why multimedia communication is so complicated.

Let us now take a look at error recovery algorithms for multimedia data streams. Until today there is no widely accepted solution to the error correction problem of multimedia streams; we discuss a number of approaches in the next chapter. An example of a new forward error correction scheme was developed at the University of Mannheim. It is called AdFEC, and we describe it briefly here.

Unlike error detection and retransmission, *forward error correction (FEC)* adds enough redundancy on the sender side to allow the receiver to reconstruct corrupted packets "on the fly." Thus, FEC adds a constant delay to all packet transmissions and thereby solves the delay jitter problem (at the cost of more redundancy). The AdFEC (Adaptable Forward Error Correction) scheme is based on binary polynomial algebra. It produces an adaptable amount of redundancy, allowing different packet types to be protected according to their importance.

Single-bit errors rarely occur in modern networks based on fiber optics. The main source of errors is packet loss in the switches. Current procedures that focus on the correction of individual bit errors do not solve this problem. Very few articles address the problem of reconstructing lost packets, mainly in the context of ATM networks [6][34][37][45]. All packets in the data stream are protected by means of the same method and with the same redundancy.

The AdFEC scheme developed at the University of Mannheim is capable of assigning different priorities to different parts of the data stream. The amount of redundancy in FEC is chosen according to the priority of a packet. A digital data stream for a movie or for audio contains more than just the digitized video/audio contents. It also contains information that must not be lost under any circumstances, such as control instructions, format data, or changes in the color lookup table. Typically, a higher error rate can be tolerated for content parts than for control parts, but all packets have to arrive on time. For example, we can assign priorities as follows:

- **Priority 1:** Packets that must not be lost under any circumstances (e.g., control and format information as well as changes in the color lookup table)
- **Priority 2:** Packets whose loss clearly adversely affects quality (e.g., audio)
- **Priority 3:** Packets whose loss is slightly damaging (e.g., pixel data in a video data stream)

For none of the three priorities is retransmission a tenable option. Starting from an already low rate of loss, third-priority packets can be transmitted without protection, second-priority packets should be protected by means of FEC with minimum redundancy, and first-priority packets by means of FEC with high redundancy.

Traditional error-correcting codes (e.g., Reed-Solomon codes) were designed for the detection and correction of bit errors. Since there is now also a demand for restoration of entire packets, new codes must be found. Specifically, errors need no longer be located—the lost bits are known. A feature of traditional error-correcting codes is their ability to locate damaged bits. The price of this feature is a great deal of redundancy. At issue here is the need to devise a code that restores the lost packets at a known error location.

Let us look at an example. Two packets a and b are to be sent. A redundancy of 100% is acceptable, i.e., up to two additional packets may be generated. These additional packets are sent together with the original packets. In the event of packet loss, the original packets a and b must be restored from the remaining packets (see Figure 5–9). In this case, two operations (labeled $\circ$ and $\bullet$) are necessary for the generation of the redundant packets.

The AdFEC algorithms described above were implemented as part of the XMovie system [31]. In the current implementation, AdFEC can generate for n given packets, n in {1,2,3}, m redundant packets, m in {1,2}. In the framework of XMovie's Movie Transmission Protocol MTP, AdFEC is used to protect parts or all of the continuous stream. The total efficiency of AdFEC is very high. AdFEC was written in C++ and ported to different UNIX workstations. Because standard C++ was used exclusively, porting of the error correction procedure to other architectures is very easy. More details can be found in [32].

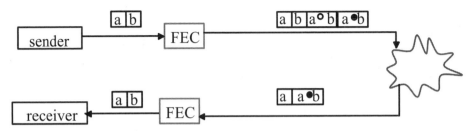

Figure 5-9 Principle of Forward Error Correction

Many other authors have also proposed the use of FEC in high-speed networks [1][6][7][34]. In conclusion, forward error correction schemes have the following advantages:

- They do not disrupt the isochronous flow of data packets.
- They are well suited for multicast environments where retransmissions are always very difficult to handle.
- They provide adaptability for different types of packets in the stream.
- They require very little computational effort to implement.

But there is also an inherent problem with FEC algorithms: If congestion of the network is the cause for packet loss, then the creation of FEC redundancy at the source will increase the number of packets in the flow and thus the loss probability. Biersack has shown that an optimal amount of redundancy can be determined [7].

An interesting example of the use of FEC is FreePhone, an adaptive audio application developed at INRIA in France. FreePhone observes the actual congestion of the network and automatically switches between different degrees of redundancy (and different audio coding techniques). In this way, the user gets the best possible audio performance for the current status of the network. The FreePhone architecture is discussed in Chapter 7.

Let us now look at our other example, the flow control algorithm. Obviously, the sliding window mechanism explained above leads to stop-and-go traffic: the packet transmitted after a stop has a much longer end-to-end delay than the packets transmitted while the window is open. The remedy is a flow control technique called "leaky bucket." Just like a bucket with a hole in the bottom, it allows a continuous stream of packets to flow out into the network. The size of the hole (i.e., the packet rate) can be adjusted by the receiver, using periodic control packets on the return path. This *rate-based flow control* is much more appropriate for continuous media streams than is window-based flow control.

This concludes our discussion of new algorithms for protocol functions. We have shown how forward error correction can solve the delay jitter problem in error handling and why rate-based flow control has advantages over window-based flow control. Fortunately, the new high-speed networks now under development all take great care to provide not only high bandwidth but also guaranteed delay bounds and jitter bounds in the lower layers [16]. At the same time, current research to extend the Internet protocol suite to enable isochronous data flows is booming [5] [53].

Now let us assume that we have replaced current end-to-end algorithms with new algorithms so that the stream of packets is isochronous from source to sink. Unfortunately, we will still not get an isochronous stream through the intermediate nodes (routers) of our packet-switched internetwork because the routers themselves will introduce a variable delay, depending on their current queue lengths and buffer allocations. These parameters are statistically distributed, and if a new high-bandwidth connection crosses a node, it will typically increase the packet delays of existing connections through that node. The problem is that the intermediate nodes do not reserve resources and thus cannot guarantee our QoS parameters. It is important to understand that Quality-of-Service mandates resource reservation and that we must add *resource reservation algorithms and protocols* to our network. Concrete protocol examples are given in section 5.3.

5.2.2 Traditional Protocols Do Not Support Multicast

As we have seen above, multicast is indeed supported in single-segment LANs and in ATM. But in an interconnected network of subnetworks, such as the Internet, this capability gets lost in layer 3. In fact, we can consider the multicast problem solved within subnetworks and must now address multicasting *between* subnetworks (i.e.,routers).

Note that multicasting is different from broadcasting: we want to reach only a subset of the nodes, not all nodes in the network. Therefore, simple algorithms such as flooding cannot be applied.

Historically, the first routing algorithms proposed for multicasting were for virtual circuit (connection-oriented) network layers; they were *sender oriented*. The assumption was that the groups of receivers are known to the sender. We can then use shortest-path algorithms, such as SPF, to compute a multicast tree for the group. If we have more than one sender within the group, a separate tree is computed for each sender; in fact, sender-oriented multicast protocols don't understand the notion of a group with multiple senders. A new receiver can only join the multicast connection by informing the sender. The sender will then re-execute the tree routing algorithm to find a new optimal multicast tree. As we will see later, the ST2 protocol and the Tenet protocol suite are examples of connection-oriented, sender-based multicast approaches.

Other approaches propose a sender-oriented multicast scheme with the possibility of new members joining and leaving the multicast tree at any time without informing the sender. The challenge is that the change in membership should not violate the QoS guarantees given to the receivers. A detailed discussion of such a scheme can be found in [20].

Current research work, especially in the Internet, centers around connectionless, *receiver-oriented* multicast, motivated by the observation that many applications will require dynamic joining and leaving of receivers, and groups with multiple senders. With the receiver-oriented paradigm, the initial sender will create a group, allocate a group (multicast) address, and start to send packets with that group address. It is now much more difficult to maintain multicast routing trees dynamically. Seminal work in this area was done by Steve Deering [11][12].

Since the initial sender does not know where potential receivers might be in the network, receiver-oriented multicast algorithms will typically begin with a broadcasting (flooding) phase, announcing the new session to all nodes, and then prune the delivery tree. Slightly more intelligent than flooding is *Reverse Path Broadcasting* (RPB). It is based on the fact that each node in the network knows from the traditional routing table its shortest path to the sender. In other words, the *inverse* paths from the receivers to the sender are well-known shortest paths, and we can take advantage of that information. The initial idea is that the RPB algorithm will forward only those packets that have arrived over the shortest path from the sender. This modification reduces the flooding of the network with duplicate packets.

Let us return to our example topology. Assume that all links have a weight of 1 and the network is in steady state. Our (still incomplete) Reverse Path Broadcasting algorithm would then create the packet flow shown in Figure 5–10 (a). We assume that C routes packets to A over link bc. Note that unlike in the case of the basic flooding algorithm, E will not send packets back to C or D.

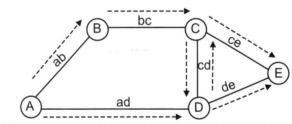

(a) The incomplete Reverse Path Broadcasting algorithm

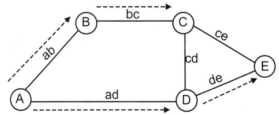

(b) The full Reverse Path Broadcasting algorithm

Figure 5–10 Reverse Path Broadcasting (RPB)

This (still incomplete) RPB algorithm has several advantages: All packets reach the receivers over the shortest paths, the overhead is reduced compared to flooding, and the additional information necessary, namely, the shortest path link towards the sender, is readily available in the nodes from the classic routing table. However, it is obvious from our example that RPB still creates unnecessary packets (two packets instead of one reach nodes C, D, and E). If we add a little extra information to the classic routing table update messages, namely, inform our neighbors what we consider our shortest path to the sender, then we can save more packets. For example, node E can now inform its neighbors C and D that de is its parent link toward A so that node C need no longer forward packets from A to E. Similarly, C and D would inform each other about their parent links to A, and no packets would be generated anymore for link cd. So, we add to our RPB algorithm the rule that a node will only forward packets to a neighbor if the link is that neighbor's parent link toward the sender. The packet flow for the complete Reverse Path Broadcasting algorithm is shown in Figure 5–10 (b). Note that the use of optimal reverse point-to-point path information has in fact created a minimal spanning tree for our multicast.

RPB will deliver the sender's packets to all nodes (i.e., all multicast routers). As the transmission of packets continues, we would like to restrict our multicast tree to those nodes that really want to participate in the group. In the receiver-oriented paradigm, the sender has no influence whatsoever on this process. We can gradually form the group of interested receiver nodes in one of two ways: either the receiving nodes send positive "membership request" packets to their parent routers or they send negative "membership denial" messages. A straightforward improvement to the Reverse Path Broadcasting mechanism is to truncate leaf nodes from the multicast tree unless they have reported their interest in the group; with this modification the algorithm is called the *Truncated Reverse Path Broadcasting* (TRPB) algorithm.

A more powerful improvement to RPB is the forwarding of denial messages, step by step, toward the root of the tree; these are then called *prune messages*. In this way, the multicast tree is pruned to those branches really necessary to reach all members of the group. An intermediate node will send a prune message over its parent link if it has no local interest in the transmission and all its children have sent prune messages to it. This algorithm is called *Reverse Path Multicasting (RPM)*.

In our example, let us assume that D and E have no interest in the group. After the initial RPB phase for the new group (Figure 5–11 (a)), node E will send a prune message to D, and D will stop forwarding packets with this group address over de (Figure 5–11 (b)). In the next phase, D will in turn send a prune message to A, and no more packets for this group will travel over link ad (Figure 5–11 (c)).

Unfortunately the Reversed Path Multicasting algorithm has three major drawbacks:

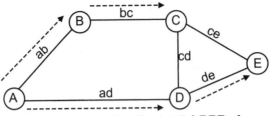

(a) Routing tree after the initial RPB phase

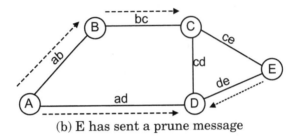

(b) E has sent a prune message

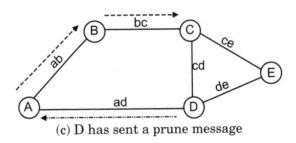

(c) D has sent a prune message

Figure 5–11 Reverse Path Multicasting (RPM)

- If a node somewhere in the pruned part of the network wants to join the group later, how can it do that? The solution is to repeat the broadcast and pruning phases periodically, adding considerable overhead to the network.
- The nodes that do not want to participate in a multicast have to pay to get out of the tree (i.e., generate prune messages). This would be similar to a telephone network where customers have to pay for each call they do *not* want to receive, an undesirable feature.
- Routing tables grow quite large since each router has to maintain either a forwarding table or prune state information for each (source, group) pair.

The PRM algorithm is used worldwide in the experimental MBone network discussed later; here, it is the basis of the *Distance Vector Multicast Routing Protocol* (DVMRP).

Before tree pruning was introduced, many local Internet operators refused to run multicast routers, and even with RPM, many of them are still reluctant to encourage MBone traffic because protocol complexity and network overhead are considerable.

As we have seen, the sender-based, connection-oriented multicast paradigm makes routing easy but has no notion of a group with multiple senders. In contrast, the receiver-based, connectionless multicast paradigm has severe routing problems: it requires periodic broadcast phases and explicit pruning messages. There are several ongoing research projects seeking a good compromise between the two.

The *Core-Based Tree (CBT)* approach proposes to establish a routing tree common to all senders of a group. One or more routers, the "core" routers, form the backbone of the tree. If a receiver wants to join the group, he sends a join packet toward the tree, and a new link is established. The messages from a sender will travel toward the core tree first, then be distributed along the branches of the tree, and finally will be sent over the leaf links to the receivers. The main advantage of CBTs is the reduction of routing table information: there is only one tree per group instead of one tree per (source, group) pair. So, CBT routing scales much better to large, multisender multicast groups than do other routing algorithms. A disadvantage is that a path from a sender through the CBT to a receiver is not necessarily the shortest path through the network, making CBT routes suboptimal. And it is not easy to find a good CBT for a given group configuration. The initial CBT idea was published in [4].

A new approach to multicast routing is taken in the *Protocol Independent Multicast (PIM)* project. Since RPM creates periodic broadcasts and lots of prune messages, it is not well suited for groups sparsely distributed over a wide area; it was initially designed for a dense population with plenty of bandwidth. There is a "dense mode" in PIM, based on periodic broadcasting, and in fact very similar to RPM. In addition, PIM has a "sparse mode," inspired by the CBT idea. Receivers send "join" messages which are propagated toward the existing distribution tree. PIM maintains a shared distribution tree for the entire group, centered at a Rendezvous Point. The Rendezvous Points all know about each other; their role in the protocol is very similar to the Core in CBTs. A Rendezvous Point is used by senders to announce their existence and by new receivers to find out about existing multicast groups.

Obviously, the nodes involved in a PIM multicast have to maintain status information about group membership. But the network layer of the Internet (the IP layer) is connectionless, so the concept of *soft state* is introduced. Soft state keeps status information within a node until a predefined time-out occurs. The state is refreshed by periodic control messages. The designers claim that soft state is more robust in the event of network failures than the "hard state" maintained by traditional connection-oriented protocols. Soft state is also the fundamental paradigm behind the resource reservation protocol RSVP, which we discuss in a later section. PIM is an ongoing research project; there is no large-scale experience yet. More information can be found in the PIM Internet draft [15].

None of the algorithms for multicast routing we have discussed so far take into account Quality-of Service requirements. In practice, a routing decision or a CBT building plan is based on very simple nearness criteria such as the number of links between two nodes (the "hop count"). Obviously, it would be very desirable to understand the QoS requirements of a data stream when making routing decisions. QoS-based routing, and its coordination with resource reservation in the nodes, is still an open research problem. An interesting algorithm for the construction of multicast trees in a network where the links have different performance characteristics is presented in [42].

5.3 A New Generation of Protocols for Multimedia

For several years there were heated debates between the "best effort" community and the "connection and reservation" community. The best-effort protagonists claimed that any kind of connection at the network layer is difficult to manage in a heterogeneous network of subnetworks; if your data stream experiences too much delay or loss, increase the bandwidth of the links and the CPU power of your routers, and if it experiences too much jitter, add a larger playout buffer at the receiver. And if you still have a problem, write "adaptive applications" that can live with changing network congestion.

The connection-and-reservation protagonists, on the other hand, believe that bandwidth and short delay will always be scarce resources, no matter how much money and infrastructure you have; QoS guarantees inherently require resource reservation, and real people will not be willing to pay real money for transmitting a multimedia data stream unless there is a firm service contract between the user and the network, specifying the stream characteristics and the QoS guarantees. As we have seen above, the best-effort philosophy dominates the Internet community; the very detailed traffic management concepts for ATM (see Chapter 4) show us clearly that the telecommunications community favors the connection-and-reservation approach. Over the last few years, both "religions" have developed their own experimental protocols. We can regard them as the forerunners of the new generation of real multimedia communication protocols, and we will thus present several examples.

5.3.1 ST2, the Stream Protocol Version II

The Internet Stream Protocol Version 2 (ST2 or ST-II) represents an early approach to multimedia communication in the Internet. ST2 is at the same level as IP and is intended to support real-time communication with multicast. ST2 had a predecessor called ST. It defined an abstraction called an "omniplex stream", where all senders of a group could send data to all receivers through the same stream. This architecture turned out to be too difficult to implement and to manage. ST2 then proposed unidirectional, connection-oriented multicast streams. The sender can define a number of stream parameters, called a *flow specification*, and the ST2 protocol establishes a multicast tree, using an ST2 implementation on each router on the path.

A revised version of the ST2 protocol has appeared as ST2+ [13]. In addition to the data transmission protocol itself, there is also a control protocol called SCMP (Stream Control Message Protocol). It is used to set up multicast connections and to reserve resources in all routers in the multicast tree. The protocol is sender oriented, i.e., the sender specifies the group of receivers for the multicast at connection setup time. The assumption is that tree routing is done only once, and the tree remains the same for the duration of the multicast connection. The tree routing algorithm is not presented in the standard, but all the message types needed to build the tree in a distributed fashion are defined.

In ST2+ it is possible to define groups of multicast trees. This feature is useful in group communication with changing senders. The observation is that network resources could be shared by such a group. For example, in an audio conference we can establish the rule that only one member of the group can send at any given time. In this case, the resources reserved for the audio streams on many of the links can be reused by several of the sender-based routing trees [13].

Each router and host implementing the ST2 protocol has a Resource Manager responsible for QoS administration. When a connection setup packet arrives at a node, the Resource Manager checks whether there are enough local resources to accommodate the new connection without violating guarantees given to existing connections (admission control). For this purpose, the full flow specification, with the stream characterization and all the required QoS, is carried in the setup packet. The ST2 standard specifies the parameters of the flow specification but gives no details on the computations to be performed for admission control (i.e., how the delays along the path have to be added up to compute the end-to-end delay). Also, it is left to the node implementor to determine how to keep the QoS promises at runtime. If the new connection can be accepted, the setup packet is passed on to the nodes downstream in the multicast tree. Acknowledgment packets travel back up the tree to the sender, establishing the final tree topology. The new connection has a globally *unique stream identifier* (SID).

In the data transfer phase, all packets generated by the sender for this stream carry the SID. The stream identifier allows all nodes along the path to treat the packets so that the QoS guarantees are met. For example, the packet scheduler within a router (see Figure 5–1) can adjust packet priorities according to the maximum delay and jitter promised to the stream. There is no error control in the data phase of ST2; if error control is desired, it must be implemented in higher layers.

Since ST2 is sender oriented, it is not easy for a new node to become a member of an existing multicast group. In the original ST2 protocol, the only way is to send a join request as a regular IP packet to the sender, and the sender can then reinitiate the multicast-tree building process, with admission control in all nodes on the new path. An advantage of sender-oriented multicast is that the sender always knows who is listening and can explicitly decide if he wants to accept a new member. This is a desirable property in many applications, such as a video conference, but undesirable in others, such as a public broadcast of an event. Note that sender-based multicast algorithms will never scale well; with a growing number of joining and leaving nodes, the sender will become a severe bottleneck. The newer ST2+ version of the protocol also allows a somewhat "against the spirit of ST2" way for receivers to join the multicast group without informing the sender.

The ST2 documents describe the packet formats and the protocol flow in detail but leave many implementation details open. Examples of undefined algorithms include multicast tree routing and the exact formulae for the computation of admission control, based on the flow specification parameters. In spite of these deficiencies, several ST2 implementations exist and are in experimental use [27][39][48].

5.3.2   The Tenet Protocols

The Tenet Group was formed in September of 1989 at the University of California at Berkeley and at the International Computer Science Institute (ICSI, also in Berkeley) to perform research in high-speed computer networking. The focus of the group is on the design and development of real-time communication services and on network support for continuous-media applications.

In its real-time communication work, the group emphasizes mathematically provable (but not necessarily deterministic) performance guarantees, contractual relationships between client and service, general parameterized user-network interfaces with multiple traffic and QoS (quality of service) bounds definable over continuous ranges, and large heterogeneous packet-switching networking environments.

The Tenet solution to guaranteed performance is based on resource reservation in all packet switches along a virtual circuit. For this purpose, the Tenet architecture proposes an extension to IP, called RTIP, and a real-time transport protocol called RMTP. Both can coexist with IP and TCP in an internetwork, as shown in Figure 5-12. The QoS parameters requested at connection establishment time, such as bandwidth, maximum end-to-end delay, and maximum delay jitter, determine the resources to be reserved in each packet switch along the path. These resources are processing power, buffer space, and "schedulability," i.e., whether the new connection can be established without violating local packet delay constraints of existing connections. The Tenet reports describe in detail, with mathematical formulae, how the required resources within a node can be determined [21][22].
The first prototype, called Tenet Scheme 1, provided real-time unicast channels only, but real-time multicast is supported in the current version, Tenet Scheme 2.

In the context of the Tenet project, an interesting mechanism for resource reservation in the connection-oriented paradigm was proposed. The idea is to initially over-reserve ("get as much as you can") on the path from the sender to the receivers when a connection is set up. If all resource requirements can be fulfilled, the connection will be accepted and the over-reservation will be relaxed by the Connect Confirm message on the way back to the sender. The total amount of

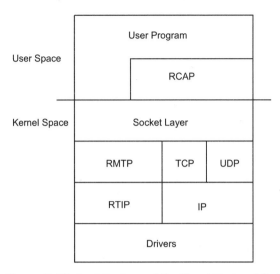

Figure 5–12 Architecture of the Tenet Protocol Suite

over-reservation will typically be evenly distributed over all intermediate nodes [23].

An open problem in current real-time multicast architectures is the dynamic joining and leaving of receivers during a connection. ST2 allows the sender to add new receivers, and the multicast tree is extended accordingly. Tenet Scheme 2 also allows dynamic group membership. The maintenance of performance guarantees for the existing and new members, multicast tree rerouting, failure management, etc., are still open problems.

5.3.3 Multicast IP and the MBone

As early as in the late 1980s, S. Deering and D. Cheriton proposed to add multicast functionality to IP in the Internet [11][12]. As a first step, the concept of multicast addresses was introduced: In addition to the traditional IP address classes A, B, and C, the new class D was defined, starting with a bit combination that identified the address field as a group address. This format is similar to IEEE addresses in LANs, where the first bit indicates whether the address is a group address or an individual station address (see Figure 5–7).

To operate in multicast mode, the IP routers have to be extended slightly to understand group addresses, duplicate incoming packets as necessary, and forward the packets on the right links. To do this most efficiently, only the packet duplication and routing functions are executed in a multicast IP router at runtime; the computation of optimal multicast routes, dynamic joining and leaving of new members, etc., are handled by a separate control protocol called IGMP (Internet Group Management Protocol).

The extended architecture of a multicast IP router is shown in Figure 5–13. Let us assume that C is on a multicast tree to D and E. Then, all incoming pack-

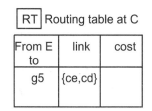

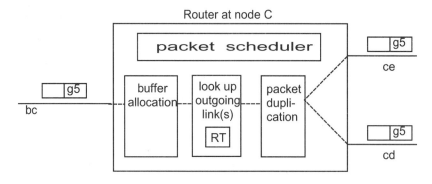

g5 = IP address of group 5

Figure 5–13 Multicast-IP-Router

ets with the group address *g5* as a destination address are duplicated, and a copy is enqueued on links *cd* and *ce*. The extensions to the IP routing function are thus relatively easy and can be implemented efficiently.

Of course, we also have to introduce multicast routing algorithms to be able to compute the paths to all receivers in a group. Multicast IP actually uses the multicast algorithms and protocols introduced above, in particular RPM with the DVMRP protocol in older versions of routers, and PIM in newer versions.

The *group management* protocols in multicast IP are not trivial, either. The basic principle is that groups can form and dissolve dynamically; with the exception of a few very special cases, group addresses are not registered with a central naming authority (unlike individual IP host addresses). This approach is reasonable if we consider video conferences, CSCW, and occasional transmissions of seminars, etc., to be the most typical multicast applications. Before a multicast is initiated at the IP level, the sender has to reserve an available group address, and the receivers have to find out what this address is. In the Internet, the Session Announcement Protocol (SAP) and the Session Description Protocol (SDP) were defined for this purpose. They inform multicast routers about existing groups so that the mapping of a group address to the packet duplication and routing functions can be performed.

Currently, IP multicast is still considered to be experimental. But since multicast IP will be fully integrated into IPv6, the next-generation IP protocol, the companion protocols will also be fully integrated into the next generation of

Internet hosts and routers. It is expected that all IPv6 products will then support IP multicast in a consistent and reliable way.

In order to gain experience with multicast IP, an overlay network for the Internet was created, called MBone (Multicast Backbone). All participating nodes are multicast IP routers; if there is an interim router on the path without IP capability, a technique called "tunneling"' is used to get to the next multicast router. Several interesting applications were written for MBone: vat (visual audio tool from Lawrence Berkeley Laboratory–LBL), vic (video conferencing tool from LBL), wb (shared whiteboard from LBL) and sdr (session directory from University College London) are the most popular ones. They can be installed on standard workstations without special hardware (except for audio I/O, and a video grabber card for the sender, of course), and are quite popular worldwide.

Experience with MBone shows several major drawbacks: The multicast tree routing algorithms do not work very well yet, and the "best effort" nature of IP, with no QoS guarantees, often leads to unsatisfactory playout of audio and video streams. It is obvious that performance guarantees cannot be given without resource reservation in the network nodes, and resource reservation requires state information, which is not available in the connectionless MBone routers.

5.3.4   IP Version 6

IP Version 6 (or IPv6 for short) is the new version of the Internet protocol IP. Currently, IP Version 4 is in use worldwide. Extensions for multimedia data streams are the main reason for the new version, but not the only one; others are a larger address space, and authentication and encryption features.

An important design goal of IPv6 is compatibility with IPv4. In a network as large and heterogeneous as the Internet, it would be impossible to migrate all IP nodes to a new version of the protocol in an atomic action. New hosts and routers running IPv6 will be able to coexist with old IPv4 hosts, enabling a smooth migration.

IPv6 is based on the same major paradigms as the older IP versions: it is connectionless (a datagram protocol), without error control or flow control in the network layer. However it has very attractive new features:

- An address space of 128 bits (rather than 32 bits) allows many more hosts to be addressed, and more levels in the address hierarchy [14].
- An improved multicast addressing scheme allows restriction of multicast routing to specified domains. The *scope* field within a multicast address limits the range of validity of an address, e.g., to an intranet within an enterprise. An additional *flags* field allows distinction between permanent and temporary multicast group addresses.
- The new *flow label* field in the header allows identification of all packets belonging to the same data stream (called a "flow" in IP). A flow is a sequence of packets sent by a host to a unicast or a multicast address. Thus,

all routers along the path can identify the packets of a flow and treat them in a flow-specific way. For example, they can schedule packets belonging to an audio stream with a higher priority than that of those belonging to a file transfer stream.

The flow label is the key feature of IPv6 for resource reservation and QoS at the IP level in the Internet. In the older versions of IP, it was not possible to identify the packets belonging to a particular multimedia stream (the source and destination addresses are obviously not sufficient), and thus resource reservation and QoS guarantees were impossible to implement.

• New mechanisms for authentication, integrity and for data encryption[2].

As we have seen in the previous section, the router uses the multicast address of an IP packet to look up an entry in the routing table; this entry determines how many duplicates of a packet will be created and on what outgoing links they will be sent. For this purpose, the current Internet Group Message Protocol (IGMP) will be integrated into the current (and quite old) Internet Control Message Protocol (ICMP); so, IPv6 will come with a new version of ICMP. For details the reader is referred to [29][47].

The header of IPv6 packets is shown in Figure 5–14. The new *flow label* field contains 28 bits, 4 bits for the *traffic class* and 24 bits for the *flow ID*. The traffic class is very similar to the traffic classes in ATM, as introduced in Chapter 4. The traffic class for a continuous stream (e.g., video or audio) will have a higher priority in a router than the traffic class for a traditional flow-controlled data stream (e.g., a TCP stream). As we have seen, the Sliding Window flow control algorithm will destroy the continuous flow anyway; thus, giving the packets of a sliding-window stream a very high priority within a router makes no sense. A router will typically discard packets of a lower-priority stream with a higher probability.

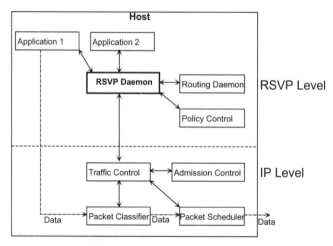

Figure 5–14 IPv6 Header

The *flow id* is a pseudo-random number generated by the source node. Together with the source address (which is, of course, also contained in the header) it forms a globally unique identifier for a flow. But how long do we maintain information about flows (e.g., scheduling priorities or resource reservations) in an IPv6 node? The protocol is connectionless, and thus we cannot discard this information at disconnect time. The proposal for IPv6 is to keep the state until a time-out is reached (i.e., we have not seen a packet on this flow for a specified time) or until a separate control protocol (in particular, RSVP) tells us explicitly to throw away the state. This concept is called *soft state*. The Internet architects claim that soft state is more robust in the event of failures and easier to manage in a very large, heterogeneous network than the "hard" state of a traditional virtual connection [14][38]. A detailed discussion of IPv6 can be found in [29].

As mentioned above, the flow label of IPv6 only makes sense in combination with a control protocol that tells the hosts and routers of a flow how to treat the packets. Unlike the case in ST2 or in the Tenet protocols, there is no connection setup packet informing the participating nodes what the characteristics of the flow will be and what its QoS requirements are.

5.3.5 RSVP, a Resource Reservation Protocol for the Internet

RSVP is currently under development as a resource reservation protocol for the Internet. After several years of discussion, the Internet community has come to accept an interesting compromise between the best-effort religion and the connection-and-reservation religion. The fundamental idea is to make resource reservations without explicitly establishing connections at the network layer.

Obviously, resource reservation only makes sense if the notion of a data stream (never call it a *connection!*) is known at the corresponding layer. As we have seen in the previous section, this will be the case in IPv6, where a stream is called a flow. So, RSVP is clearly meant to be a companion protocol to IPv6.

Unlike ST2 or the Tenet protocols, RSVP is *receiver oriented*. The receiver sends out a reservation message with a flow specification. He identifies the data stream he wants to see, and the message is forwarded toward the sender. The sender does not know the set of group members currently receiving his multicast stream. This makes dynamic joining and leaving much easier than a sender-oriented scheme, at the cost of sender control over the participants [53].

To determine QoS requirements, the receiver uses both local data about available resources (e.g., for the playout of a data stream) and data arriving from the sender. A technique *called One-Pass-with-Advertising* (OPWA) allows the packets initiated by the sender to collect information about resources within the network on the path to the receiver and to deliver those as *advertisements* to the receiver, together with a traffic characterization (SenderTspec). Based on these, the receiver can determine the resources he wishes to reserve on intermediate nodes [43].

Similarly to ST2 and the Tenet protocols, each router running RSVP must be able to manage resources explicitly. This function is called *Traffic Control*, and it consists of three parts: admission control, packet classifier, and packet scheduler. Upon arrival of a reservation request, admission control runs a local algorithm to determine the availability of the requested resources. If they cannot be granted, the sender of the request is informed accordingly. Otherwise, the local packet classifier and packet scheduler are instructed to accommodate the new flow requirements, and the reservation request is forwarded to the next router.

At runtime, an incoming data packet is first passed to the packet classifier. It determines the route and the local QoS for the packet; in essence, it maps the parameters of the incoming packet to a locally known QoS class. The classifier can use any combination of parameters from the incoming packet, e.g., pairs of source address/source port and destination address/destination port in IPv4, or a flow label in IPv6. The packet scheduler then determines the priority of the packet and forwards it to the appropriate output queue. If the lower-layer subnetwork understands resource reservation, then those reservations are also handled by the packet scheduler.

The architecture of a router equipped with RSVP is shown in Figure 5–15. Clearly, it is meant to be a straightforward and efficient extension to a multicast IP router. The exact details of an implementation depend on the resources available locally (buffers, CPU cycles, etc.) and are not described in the RSVP documents (just as in ST2). In particular, the formulae for admission control are not easy to implement in practice: For a given set of existing flows, can the router guarantee, under all circumstances and with the remaining resources, the requested QoS for this new reservation?

In RSVP a reservation request is described in the form of a flow descriptor. It consists of a FlowSpec and a FilterSpec. The exact formats and contents are transparent for the RSVP protocol; it only delivers these data fields to the routers for further processing.

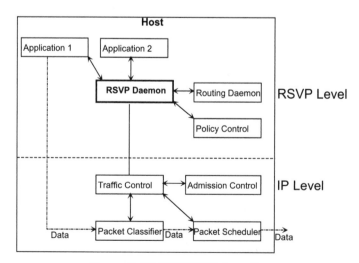

Figure 5–15 RSVP/IP Interaction

In order to support a wide variety of applications and thus multicast topologies, RSVP allows very flexible grouping of reservations. In a first option, a receiver can specify in his request whether the reservation should be *distinct* (i.e., only for packets coming from one particular sender) or *shared* (i.e., for the accumulated packet streams coming from all senders). In a second option, he can restrict the reservation to a specified list of senders (*explicit*) or allow all senders who might ever send data to use the resources (*wildcard*). The combination of these options is shown in Figure 5–16 (see [9]). For example, in a video conference scenario, a receiver would use the WF style of reservation for an audio stream; anyone sending audio to the group would have to use the same resources, under the assumption that only one person would speak at a time. He might use the FF style, with one reservation per sender, to ensure that a separate, high-quality video window would show every single participant in the conference.

Reservation ⟍ ⟍ Choice of Sender ⟍	Distinct	Shared
Explicit	Fixed-Filter (FF)-Style	Shared-Explicit (SE) Style
Wildcard	Not defined	Wildcard Filter (WF-Style)

Figure 5–16 RSVP Reservation Options

Similar to multicast IP, the RSVP protocol also allows *tunneling*: If an intermediate IP node between two RSVP nodes does not implement RSVP, it can still forward the RSVP messages. However, no resources will be reserved on the intermediate node; thus, the end-to-end path will have a best-effort link, and deterministic end-to-end guarantees will no longer be possible.

Obviously, the reservation information to be maintained in a node is state information. But RSVP is connectionless, so how long will a node store the state? The solution is to implement the *soft state* concept introduced above (similar to IPv6): state information times out and is automatically dropped from a node unless it is refreshed regularly. Refreshing can be done by both the receivers (by RESV messages) and the senders (by PATH messages). For details the reader is referred to the RSVP document [9].

An architectural comparison between RSVP and ST2 can be found in [36]. Early implementations of RSVP are described in [3][49]. No large-scale experiments in a complex internetwork have been reported yet.

5.3.6   RTP, a Real-Time Transport Protocol

RTP is a transport protocol for multimedia data streams in the Internet. It was designed to run over multicast IP (in fact, RTP packets are typically sent over UDP, which provides an application programming interface for IP), in order to provide timing information and stream synchronization. It is a light-weight protocol, without error correction or flow control functionality. In principle, RTP could also be operated over other protocols. The requirements for the underlying lower-layer protocol are minimal.

The format of the RTP header is show in Figure 5–17. The timestamp field is used for both intrastream and interstream synchronization. It records the creation time of the first byte of the packet. The timestamp resolution depends on the type of data stream; for example, a timestamping frequency of 65,536 Hz is often used for digital video, whereas audio streams are timestamped at the sampling rate. Several subsequent RTP packets can bear the same timestamp if they belong to the same application data unit of the stream, i.e., one video frame. The timestamp is used by the receiver to maintain synchronization with real-time or with other multimedia streams. The Synchronization Source identifier (SSRC) is a random number, generated by the sender, and unique for the lifetime of an RTP session. Colliding SSRC identifiers are detected and resolved. The SSRC allows a receiver to uniquely identify a data stream.

RTP introduces two types of intermediate nodes between senders and receivers: mixers and translators. A *mixer* receives RTP packets from one or several senders, combines them, and forwards them in a new RTP packet. The stream of combined packets will have a new SSRC ID. The SSRC IDs of contributing senders will be added to the packets as Contributing Sources (CSRCs). Because packets from different contributing senders can arrive out of synch (they can travel along different paths through the network), the mixer changes the temporal structure of the streams. An example of a mixer is the combination of multiple audio sources of a multipeer conference into one audio stream, to be forwarded to all receivers. In contrast to a mixer, a *translator* modifies only packet contents, without combining streams. An example would be a video coding converter or a firewall filter.

RTP also comes with a control protocol, RTCP. It sends transmission reports to all participants in order to monitor the performance and quality of the streams. It is also used by the senders to ensure the uniqueness of SSRC identifiers.

A detailed explanation of RTP and RTCP can be found in the Internet RFC [44]. In spite of the fact that RTP is a very young protocol, a number of Internet applications already implement RTP, in particular, the MBone tools vic and vat. The RTP algorithms are usually not implemented as a separate layer but are part of the application code. It is expected that next-generation WWW browsers will also use RTP for live video and audio streams.

The reader should be aware that RTP provides neither resource reservation nor QoS support. It adds a relatively small but important functionality to con-

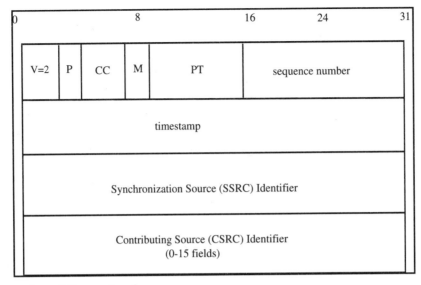

0			8		16	24	31

V=2	P	CC	M	PT	sequence number

timestamp

Synchronization Source (SSRC) Identifier

Contributing Source (CSRC) Identifier
(0-15 fields)

V = version of the protocol
P = padding on/off (if on, there will be padding bytes at the end of the packet)
CC = CSRC Counter (number of CSRC fields contained in the packet)
M = Mark (profile dependent)
PT = payload type (type of payload; defined in a companion RFC)

Figure 5–17 RTP Header Format

nectionless, best-effort protocols: namely, timing. The overhead is minimal and the protocol is very efficient.

5.4 Media Filtering, Media Scaling, and Adaptive Applications

Most of the algorithms we have seen so far assume that an application knows its QoS requirements in advance. When a connection is set up, or (in IP terms) a flow is initialized, the resources required to maintain the quality of service are reserved along the path from the sender to the receivers. The network guarantees that these resources will be available throughout the duration of the connection.

However, in a multicast scenario, not all receivers have the same QoS requirements. For example, a PC connected via a telephone line will not be able to receive video at the same rate as a high-end UNIX workstation connected via ATM. A solution to this problem is *media filtering*.

5.4.1   Media Filtering

The basic idea is to add more intelligence to the internal network nodes; if they understand the contents of a data stream, they can filter out packets on their way down the multicast tree.

Figure 5–18 shows an example. The leaf nodes are the receivers of a video multicast. The different receivers have different QoS requirements based on their local processing power, link capacity, and end-user preferences. If the network is unable to filter the media stream, the sender has to set up four separate connections, wasting bandwidth on the links close to the root of the tree. On the other hand, if the internal network nodes implement media filters, the sender will create only one flow satisfying the maximum QoS, and considerable bandwidth will be saved.

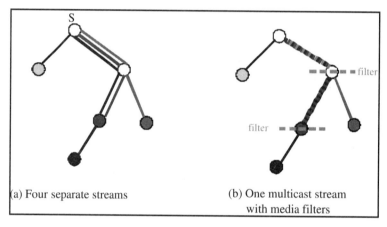

(a) Four separate streams   (b) One multicast stream
                                             with media filters

Figure 5–18   Media Filtering

Media filtering can work only if a multimedia data stream can be decomposed meaningfully into "weaker" substreams. For example, a PCM-encoded audio stream cannot be reduced by omitting three out of four packets. *Hierarchical stream encoding* is required. If we recall the video compression techniques presented in Chapter 3, we now understand why MPEG-1 and MPEG-2 have a hierarchical encoding mode: An MPEG video stream can be packetized in such a way that by leaving out m packets from a frame consisting of n packets, we obtain a complete image with lower quality. This is sometimes called Layered Receiver-Driven Multicast (LRM). We now understand the close relationship between media encoding techniques and algorithms and protocols in the network, and that is why Chapter 3 is part of this book. For details, the reader is referred to the literature [46][51][52].

How much intelligence should be built into the network nodes is still an open research question. Some researchers argue the internal nodes of a network should be as efficient and application independent as possible; the explosive growth of the Internet is largely due to this design principle. Others even propose the remote loading of arbitrary, user-written filter programs into the internal nodes of the network; this would offer maximum flexibility but introduce the danger of Trojan Horses. Many network operators would probably refuse to download user-written code into their network nodes. A reasonable intermediate solution would be to write parameter-controlled filters for a small set of widely accepted stream encodings and provide these as part of the code running in the network nodes, similar to plug-ins for WWW browsers. When a stream is set up, the parameters of the filters are set by the receivers or the sender to optimize media filtering. The RSVP protocol is perfectly suited to this approach.

5.4.2 Media Scaling

Another problem with a static "QoS contract" between the senders and receivers of a multicast stream is the variance of many parameters throughout the duration of the transmission, both at the end nodes and within the network. It would be desirable to *adjust* QoS parameters during a multimedia connection. When applied to a multimedia data stream, this is called *media scaling*.

The traditional connection-oriented protocols have always supported a simple method of media scaling, namely, flow control, preventing a fast sender from swamping a slow receiver. Both the window-based and rate-based flow control schemes discussed above adjust the packet rate of the sender according to the current status of the receiver and the network. Note that the only parameter controlled by these algorithms is the data rate. This is reasonable for non-real-time connections (e.g., for e-mail, file transfer, news transfer, etc.), but it obviously spoils the isochronous transfer of real-time streams.

Media scaling allows control of parameters other than just the data rate. In a video stream, the image quality is an obvious candidate. The amount of redundancy in a forward error correction scheme (see section 5.2.1) would be another candidate: the more redundant data we transmit, the higher the probability that packet losses can be hidden from the user at the receiver's site. To implement media scaling, the interface between the application and the network must be extended to pass control information up and down. For example, if the network signals increasing congestion, an MPEG video encoder could adjust the quantization stepsize (see Chapter 3): the fewer quantization intervals we have, the fewer bits we need per pixel block, and the lower will be the data rate and the image quality.

Many research projects address the issue of media scaling—see, for example, [30][35][50]—but there are no standards yet for the application/network interface or for the algorithms and protocols within the network.

5.4.3 Adaptive Applications

Some researchers propose a very radical approach to QoS and media scaling: they oppose any kind of resource reservation within the network and argue that applications should be able to adapt to arbitrary changes in the network conditions [8][18]. Their main arguments are:

- Resource reservation is very costly in terms of protocol messages.
- Over-reservation leads to an inefficient use of resources.
- Firm reservations cannot adapt to streams with QoS variations during their lifetime.
- If the network really gets too congested, you can always throw bandwidth at the problem.

So, the idea is to run multimedia streams over a traditional best-effort network, such as the Internet. The senders can vary their packet rates at will. If the network signals congestion, the applications will have specific mechanisms with which to react, such as the modification of the quantization parameter in MPEG or the choice of another audio compression algorithm. Adaptive applications use media scaling as their only means to adjust QoS; there is no reservation protocol or QoS contract with the network.

Whether or not adaptive applications over best-effort networks are the right solution to multimedia communication is a subject of heated debate. It is probably a quick and reasonable solution if the entire network is under the control of one organization. But interoperation with service providers in a worldwide network and cross-traffic from foreign connections that which would not agree to being scaled down will probably require some kind of resource reservation and QoS guarantees.

5.5 Summary

This chapter began with a closer look at the main functions of traditional protocols in layers 3 and 4: packet switching and packet routing, and reliable transfer. We introduced the most important routing algorithms in use today, and the error detection and sliding window flow control mechanisms implemented in all major transport protocols. We then gave an overview of multicast support in layer 2 and concluded that an additional multicast function is required at layer 3 for internetworks.

In the next section, we tried to match the multimedia communication requirements with the traditional protocol functions. We showed that traditional protocols cannot meet QoS requirements; in fact, many algorithms spoil the isochronous flow of packets. Resource reservation is not supported; the traditional networks are best-effort. We introduced forward error correction and rate-based flow control and claimed that these algorithms are better suited for continuous streams. Also, traditional networks do not support multicast in layer 3. We have presented a new generation of routing algorithms, developed mainly in the context of the Internet, which are able to establish multicast trees for packet-switched networks.

After explaining the basic principles of the new algorithms, we proceeded with concrete examples of new protocols for multimedia, developed over recent years in research projects. These projects centered around different paradigms: some argue for sender-oriented, one-to-many multicast streams; others favor receiver-oriented, many-to-many group protocols. Some schemes are connection-oriented (ST2, Tenet), others are connectionless and/or soft-state (multicast IP, IPv6, RSVP). Some are still best-effort (multicast IP), others incorporate QoS parameters and resource reservation (ST2, Tenet, RSVP).

In the final, major section, we presented the idea of adapting the media streams to varying user requirements and changing network conditions. This adaptation comes in three flavors: media filtering, media scaling, and adaptive applications. All of them are based on the concept of hierarchical encoding so that the "quality" of a stream can be adapted to varying resources.

If readers have gotten the impression that every one of those approaches concentrates on some of the issues of multimedia communication but there is no clear overall architecture and matching set of protocols, then they are perfectly right.

References

[1]   A. Albanese and M. Luby, "PET – Priority Encoding Transmission," In *High-Speed Networking for Multimedia Applications*, edited by W. Effels-berg, O. Spaniol, A. Danthine, and D. Ferrari. Dordrecht: Kluwer Academic Publishers, 1996:247–265.

[2]   R. Atkinson, *Security Architecture for the Internet Protocol*, RFC 1815, Network Working Group, Internet Engineering Task Force (Internet Engineering Task Force (IETF)), August 1995.

[3]   S. Berson, *ReSerVation Protocol Daemon for SunOS/BSD*, University of Southern California Information Sciences Institute, January 1995.

[4]   T. Ballardie, P. Francis, and J. Crowcroft, "Core based Trees - An Architecture for Scalable Inter-Domain Multicast Routing," *Proc. ACM SIGCOMM* 1993:85-95.

[5]   A. Banerjea, D. Ferrari, B. Mah, and M. Moran, *The Tenet Real-Time Protocol Suite: Design, Implementation and Experiences.* ICSI TR-94-059, International Computer Science Institute, Berkeley, 1994, available via ftp or WWW from icsi.berkeley.edu

[6]   E. W. Biersack, "Performance Evaluation of Forward Error Correction in ATM Networks," *Computer Communication Review* 22(4):248–257, 1992.

[7]   E. W. Biersack, "Performance Evaluation of Forward Error Correction in an ATM Environment," *IEEE Journal on Selected Areas in Communications* 11(4):631–640, 1993.

[8]   J.-C. Bolot, T. Turletti, and I. Wakeman: "Scalable Feedback Control for Multicast Video Distribution in the Internet," *Proc. ACM SIGCOMM* 1994:58–67.

[9]   B. Braden, L. Zhang, D. Estrin, S. Herzog, and S. Jamin, *Resource Reservation Protocol (RSVP), Version 1 Functional Specification*, Internet Draft, Internet Engineering Task Force (IETF), August 1996.

[10] C. Cheng, R. Riley, S. Kumar, and J. J. Garcia-Luna-Aceves, "A Loop-free Extended Bellman-Ford Routing Protocol Without Bouncing Effect," *Proc. ACM SIGCOMM* 1989:224–236.

[11] S. E. Deering and D. R. Cheriton, *Host Groups: A Multicast Extension to the Internet Protocol*, Internet RFC 966, December 1985.

[12] S. E. Deering and D. R. Cheriton, "Multicast Routing in Datagram Internetworks and Extended LANs," *ACM Trans. on Computer Systems* 8(2): 85–110, 1990.

[13] L. Delgrossi and L. Berger, *Internet Stream Protocol Version 2+*, Internet Engineering Task Force (IETF), 1995

[14] S. Deering and R. Hinden, *IPv6 Addressing Architecture*. Network Working Group, Internet Engineering Task Force (IETF), December 1995.

[15] S. Deering, D. Estrin, D. Farinacci, V. Jacobson, C. Liu, and L. Wei, Protocol Independent Multicast (PIM): Motivation and Architecture. Internet Draft, URL: http://netweb.usc.edu/lwei/arch/PIM-arch-ID.txt

[16] M. de Prycker, *Asynchronous Transfer Mode*. Prentice Hall Europe, 1995.

[17] E.W. Dijkstra, "A Note on Two Problems in Connection With Graphs," *Numerische Mathematik* 1:269-271, 1959.

[18] Ch. Diot, "Adaptive Applications and QoS Guarantees," *Proc. IEEE Internat'l. Conf. on Multimedia and Networking*, Aizu, Japan, 1995, IEEE Computer Society Press, 1995:99-106.

[19] M. Doar, *Multicast in the Asynchronous Transfer Mode Environment*. PhD dissertation, University of Cambridge, 1993.

[20] W. Effelsberg and E. Müller-Menrad, *Dynamic Join and Leave for Real-Time Multicast*. International Computer Science Institute, Technical Report TR-93-056. Available via ftp or WWW from icsi.berkeley.edu.

[21] D. Ferrari, "Realtime Communication in an Internetwork," *Journal of High-Speed Networks* 1(1):79–103, 1992.

[22] D. Ferrari, "A New Admission Control Method for Real-Time Communication in an Internetwork," In *Principles of Real-Time Systems*, edited by S. H. Son. Englewood Cliffs, NJ: Prentice Hall, 1994.

[23] D. Ferrari and D. C. Verma, "A Scheme for Real-Time Channel Establishment in Wide-Area Networks," *IEEE Journal On Selected Areas In Communications* 8(4):368–379, 1990.

[24] F. Fluckiger, *Understanding Networked Multimedia: Applications and Technology*. Upper Saddle River, NJ: Prentice-Hall, 1995.

[25] J. J. Garcia-Luna-Aceves, "A Unified Approach to Loop-Free Routing using Distance Vectors or Link States," *Proc. ACM SIGCOMM 89*, Austin, Texas, 1989:212-223.

[26] J. J. Garcia-Luna-Aceves and S. Murthy, "A Path-Finding Algorithm for Loop-Free Routing," *IEEE/ACM Transactions on Networking* 5(1):148–160, 1997.

[27] R.G. Herrtwich, "An Architecture for Multimedia Data Stream Handling and Its Implication for Multimedia Transport Service Interfaces," *Proc. Third Workshop on Future Trends of Distributed Computing Systems*, Taipei, Taiwan, April 1992, IEEE Computer Society Press, 1992:269–75.

[28] Ch. Huitema, *Routing in the Internet*. Upper Saddle River, NJ: Prentice Hall, 1995.

[29] Ch. Huitema, *Ipv6 – The New Internet Protocol*. Upper Saddle River, NJ: Prentice Hall, 1996

[30] T. Käppner and L. C. Wolf, "Media Scaling in Distributed Multimedia Object Services," In *Multimedia: Advanced Teleservices and High-Speed Communication Architectures*, edited by R. Steinmetz. Springer LNCS 868, 1994:34–43.

[31] R. Keller, W. Effelsberg, and B. Lamparter, "XMovie: Architecture and Implementation of a Distributed Movie System," *ACM Transactions on Office Information Systems* 13(4):471–495, 1995.

[32] B. Lamparter, O. Böhrer, V. Turau, and W. Effelsberg: *Adaptable Forward Error Correction for Multimedia Data Streams*. TR 93-009, Praktische Informatik IV, University of Mannheim, 1993, URL: http://www.informatik.uni-mannheim.de/informatik/publications/index.publications.html

[33] S. Lin and J. Costello, *Error Control Coding*. Englewood Cliffs: Prentice Hall, 1983

[34] A. J. McAuley, "Reliable Broadband Communication Using a Burst Erasure Correcting Code," *Proc. ACM SIGCOMM '90 Symposium. Communications Architectures and Protocols, ACM Computer Communication Review* 20(4):297–306, 1990.

[35] S. McCanne, *Scalable Compression and Transmission of Internet Multicast Video*. Ph.D. dissertation, University of California, Berkeley, 1996.

[36] D.J. Mitzel, D. Estrin, S. Shenker, and L. Zhang, "An Architectural Comparison of ST-II and RSVP," *Proc. IEEE INFOCOM*, 1994.

[37]   H. Ohta and T. Kitami, "A Cell Loss Recovery Method Using FEC in ATM Networks," *IEEE Journal on Selected Areas in Communications* 9(9):1471-1483, 1991.

[38]   C. Partridge: *Using the Flow ID field in IPv6*, RFC 1809, Network Working Group, Internet Engineering Task Force (IETF), June 1995.

[39]   C. Partridge and S. Pink, "An Implementation of the Revised Internet Stream Protocol (ST-2)," *Internetworking: Research and Experience* 3(1):27-54, 1992.

[40]   B. Plattner and C. Lanz, *X.400 Message Handling: Standards, Internetworking, Applications*. Reading, MA: Addison Wesley, 1991.

[41]   R. C. Prim, "Shortest Connection Networks and Some Generalizations," *Bell System Technical Journal* 36:1389-1401, 1957.

[42]   S. Ramanathan, "Multicast Tree Generation in Networks with Asymmetric Links," *IEEE/ACM Transactions on Networking* 4(4):558-568, 1996.

[43]   S. Shenker and L. Breslau, "Two Issues in Reservation Establishment," *Proc. ACM SIGCOMM*, Cambridge, MA, 1995:14-26.

[44]   H. Schulzrinne, S. Casner, R. Frederick, and V. Jacobson, *RTP: A Transport Protocol for Real-Time Applications*, RFC 1889, Internet Engineering Task Force (IETF), January 1996.

[45]   N. Shacham and P. McKenny, "Packet Recovery in High-Speed Networks using Coding," *Proc. INFOCOM 90*, San Francisco,1990.

[46]   R. Steinmetz and K. Naehrstedt, *Multimedia: Computing, Communications and Applications*. Upper Saddle River, NJ: Prentice Hall, 1995.

[47]   A. Tanenbaum, *Computer Networks*, 3rd ed. Upper Saddle River, NJ: Prentice Hall, 1996.

[48]   C. Topolcic, *Experimental Internet Stream Protocol, Version 2 (ST-II)*. Internet RFC 1190, October 1990.

[49]   A. Urban, *Untersuchung der Ressourcenreservierung in Rechnernetzen am Beispiel von RSVP*. Diplomarbeit (Master's thesis), Praktische Informatik IV, University of Mannheim, Germany, 1996 (in German).

[50]   H. Wittig, J. Winckler, and J. Sandvoss, "Network Layer Scaling: Congestion Control in Multimedia Communication with Heterogeneous Networks and Receivers," *Proc. Multimedia Transport and Teleservices* (1994), Springer LNCS 882, pp. 274-293.

[51]   L. Wolf, R. G. Herrtwich, and L. Delgrossi: "Filtering Multimedia Data in Reservation-based Networks," *Proc. Kommunikation in verteilten Systemen*, Springer-Verlag 1995:101-112.

[52] N. Yeadon, A. Mauthe, D. Hutchison, and F. Garcia, "QoS Filters: Addressing the Heterogeneity Gap," In *Proc. Interactive Distributed Multimedia Systems and Services*, edited by B. Butscher, E. Müller, and H. Pusch. Berlin: Springer LNCS 1045, 1996:227-244.

[53] L. Zhang, S. Deering, D. Estrin, S. Shenker, and D. Zappala, "RSVP: A New ReSerVation Protocol," *IEEE Network*, September 1993.

CHAPTER 6

End-to-End Reliable Multicast

J. J. Garcia-Luna-Aceves
Brian Neil Levine

*I*n this chapter, we discuss emerging end-to-end reliable multicast protocols. The importance of these protocols is a consequence of the reliable-transmission service requirements of new, interactive, distributed, multimedia applications running over long-haul networks and the Internet. For brevity, our treatment concentrates on protocols that would operate over the IP Internet. We review known classes of reliable multicast protocols for long-haul networks and internetworks, compare the relative performance of these classes using an approximate maximum throughput model, discuss issues associated with the implementation of real protocols that constitute practical instantiations of the protocol classes, and outline emerging solutions for the scaling problems we identify in reliable multicast protocols.

6.1 Defining End-to-End Reliability

Such interactive, distributed multimedia applications as shared whiteboards, group editors, and distributed simulations over long-haul networks and the Internet require end-to-end reliable and unreliable multicast services. This chapter focuses on end-to-end reliable multicast protocols.

Although protocols for reliable multicasting over local area networks (LANs) have been around for quite some time [3], architectures and protocols for the provision of end-to-end reliable multicasting (reliable multicasting, for short) over the Internet are still evolving. The design problems of such protocols are compounded by the Internet's size and its explosive growth, the variety of applications that require reliable multicasting, and the varying sizes and dynamics of the user groups within which reliable multicasting must take place.

The primary functions of a reliable multicast protocol are to provide a reliable transmission service for each multipoint conversation or multicast session under way and to multiplex multiple conversations over the same access points to the network or internetwork. As is the case in point-to-point transport protocols [2], additional functions of reliable multicast protocols can include security, flow control, and packetizing. Typically, a reliable multicast protocol assumes the existence of multicast routing tree(s) provided by underlying multicast routing protocols. In the Internet, these routing trees can be built by using such protocols as Distance Vector Multicast Routing Protocol (DVMRP) [5][6], Core Based Trees (CBT) [1], Ordered Core Based Trees (OCBT) [21], Protocol Independent Multicast (PIM) [7], and the Multicast Internet Protocol (MIP) [15].

The semantics of the end-to-end transmission service provided by a reliable multicast protocol depends on the service requirements of the applications making use of the protocol. In the case of unicast transport protocols, two basic semantics have been implemented widely: unreliable end-to-end transmission (e.g., UDP); and end-to-end transmission services that pass packets in the correct sequence, with no duplicates, and no omissions (e.g., TCP). In contrast, the range of reliable multicast services that applications may require is much more varied. For example, the Scalable Reliable Multicast (SRM) protocol [8] assumes that the applications are capable of rebuilding the information needed for retransmission in an entire session; accordingly, although SRM retransmits packets after perceived errors or losses, it does not guarantee reliable delivery of all data. At another extreme, the application may require the underlying reliable multicast protocol to ensure that the receivers of a multicast session receive the information sent by all sources free of errors, with no duplicates, and preserving causal order among all data communicated among participants.

Regardless of the type of reliability required by the application, a reliable multicast protocol operating over the Internet must employ mechanisms to handle occasional errors, losses, duplications, and reordering of packets in transit, and these mechanisms rely on feedback from the receivers to the senders. As the number of receivers in a multicast group increases, the sources of data are forced to process more and more messages from the receivers that either indicate correct data reception or request retransmissions; this problem is commonly called the *acknowledgment (ACK) implosion* problem.

6.2 A Taxonomy of Reliable Multicast Protocols

As we ointed out in the previous section, there can be many service definitions for end-to-end reliable multicast protocols. We focus on what could be considered the minimum end-to-end reliable service that can be provided to applications, which consists of ensuring that all receivers obtain the packets sent by a given source within a finite time. We classify many of the reliable multicast protocols already implemented or proposed. These protocols are categorized according to the mechanisms used to pace the transmission of new packets and the mechanisms used to allow sources to delete transmitted packets within a finite time in the course of a session. Withthis approach, the protocol can be thought of as using two windows: a *congestion window* (cw), which that advances based on feedback from receivers regarding the pacing of transmissions and detection of errors, and a *memory allocation window* (mw), which advances based on feedback from receivers as to whether the sender can erase data from memory. In practice, protocols may use a single window for pacing and memory (e.g., TCP [11]) or separate windows (e.g., NETBLT [4]).

6.2.1 Sender-Initiated Protocols

A sender-initiated protocol requires the source to maintain the state of all the receivers to whom it has to send information and from whom it has to receive feedback. In the past [18][19], sender-initiated protocols have been described as placing the responsibility of reliable delivery at the sender. However, this characterization is overly restrictive and does not reflect the way in which several reliable multicast protocols that rely on positive acknowledgments from the receivers to the source have been designed. In our taxonomy, a sender-initiated protocol is one that requires the source to receive acknowledgments from all the receivers before it is allowed to release memory for the data associated with the acknowledgments. Accordingly, the source is required to know the constituency of the receiver set, and the scheme suffers from the ACK-implosion problem. Either the source or the receivers can be in charge of the retransmission time-outs used for pacing of transmissions.

The traditional approach to pacing and transmission error detection (e.g., TCP in the context of reliable unicasting) is for the source to be in charge of the retransmission time-out. However, as suggested by the results reported by Floyd et al. [8], a better approach for pacing a multicast session is for each receiver to set its own time-out. A receiver sends ACKs to the source at a rate that it can accept and sends a negative acknowledgment (NACK) to the source after not receiving a correct packet from the source for an amount of time that exceeds its retransmission time-out. An ACK can refer to a specific packet or a window of packets, depending on the specific retransmission strategy. A simple illustration of a sender-initiated protocol is presented in Figure 6–1. Notice that, regardless of whether a sender-based or receiver-based retransmission strategy is used, the source is still in charge of deallocating memory after receiving all the ACKs for a given packet or set of packets. The source keeps packets in memory until every receiver node has acknowledged receipt of the data. If a sender-based retransmission strategy is used, the sender "polls" the receivers for ACKs by retransmitting after a time-out. If a receiver-based retransmission strategy is used, the receivers "poll" the source (with a NACK) after they time out. Of course, the source still needs a timer to ascertain when its connection with a receiver has failed.

Figure 6–1 Sender-Initiated Protocols

It is important to note that just because a reliable multicast protocol uses NACKs, it does not mean that they are the basis for the source to ascertain when it can release data from memory. The combination of ACKs and NACKs has been used extensively in the past for reliable unicast and multicast protocols. For example, NETBLT [4] is a unicast protocol that uses a NACK scheme for retransmission, but only on small partitions of the data. In between the partitions, called "buffers," are ACKs for all the data in the buffer. Only upon receipt of this ACK does the source release data from memory; therefore, NETBLT is sender-initiated. In fact, NACKs are unnecessary in NETBLT for its correctness, i.e., a buffer can be considered one large packet that eventually must be acknowledged, and are important only as a mechanism to improve throughput by allowing the source to know sooner when it should retransmit some data.

A protocol similar to NETBLT is the "Negative Acknowledgments with Periodic Polling" (NAPP) protocol [20]. This protocol is a broadcast protocol for LANs. Like NETBLT, NAPP groups together large partitions of the data that are periodically acknowledged, while NACKs are used to indicate lost packets within the partition. Because the use of NACKs can cause an implosion problem at the source, NAPP uses a NACK-avoidance scheme. As in NETBLT, NACKs increase NAPP's throughput but are not necessary for its correct operation, albeit slow. The use of periodic polling limits NAPP to LANs because the source can still suffer from an ACK-implosion problem even if ACKs occur less often.

Other sender-initiated protocols, like the Xpress Transfer Protocol (XTP) [22], were created for use on an internet, but they still suffer from the ACK-implosion problem.

The main limitation of sender-initiated protocols is not that ACKs are used, but that the source must process all of the ACKs and must know the receiver set. The two known methods that address this limitation are (a) using NACKs instead of ACKs, and (b) delegating retransmission responsibility to members of the receiver set by organizing the receivers into a ring or a tree. We discuss both approaches subsequently.

6.2.2 Receiver-Initiated Protocols

Previous work [18][19] characterizes receiver-initiated protocols as placing the responsibility for ensuring reliable packet delivery at each receiver. The critical aspect of these protocols for our taxonomy is that no ACKs are used to let the sources know when all the receivers have correctly obtained a given sequence of packets. The receivers send NACKs back to the source when a retransmission is needed, as detected by an error, a skip in the sequence numbers used, or a time-out. Because the source receives feedback from receivers only when packets are lost and not when they are delivered, the source is unable to ascertain when it can safely release data from memory. There is no explicit mechanism in the ideal receiver-initiated protocol for the source to release data from memory, even though its pacing and retransmission mechanisms are scalable and efficient. Figure 6–2 is a simple illustration of a receiver-initiated protocol.

Figure 6–2 Receiver-Initiated Protocols

Receiver-initiated protocols can experience a NACK-implosion problem at the source if many receivers detect transmission errors. To remedy this problem, a subclass of receiver-initiated protocols [8][18][19] adopts the NACK-avoidance scheme first proposed for NAPP; we call this type of protocols *receiver-initiated with NACK-avoidance* (RINA) protocols. The generic RINA protocol has been shown [18][19] to improve substantially on the performance of the basic receiver-initiated protocol.

The basic operation of a RINA protocol is as follows [18][19]: The sender multicasts all packets and state information, giving priority to retransmissions. Whenever a receiver detects a packet loss, it waits for a random time period and then multicasts a NACK to the sender and all other receivers. When a receiver obtains a NACK for a packet that it has not received and for which it has started a timer to send a NACK, the receiver sets a timer and behaves as if it had sent a NACK. The expiration of a timer without the reception of the corresponding packet is the signal used to detect a lost packet. With this scheme, it is hoped that only one NACK is sent back to the source for a lost transmission for an entire receiver set. Nodes farther away from the source might not even get a chance to request a retransmission. The generic protocol does not specify how timers are set accurately.

The main motivation for RINA protocols is that using NACKs frees the sender from having to process every ACK from each receiver. Two additional advantages are that the source is not supposed to know the receiver set and that the receivers pace the source. However, because there is no mechanism in the basic RINA protocol for the source to know when it can safely release data from memory [12], the data needed for retransmission must be rebuilt by the application, and all data exchanged in a session must be kept by the sources or some archival server at least for the length of the session.

An example of a receiver-initiated protocol is the *"log-based receiver-reliable multicast"* (LBRM) [10], which uses a hierarchy of log servers that store information indefinitely and in which receivers recover by contacting a log server. Using log servers is feasible only for applications that can afford the servers and leaves many issues unresolved. If a single server is used, performance can degrade due to the load at the server; if multiple servers are used, mechanisms must still be implemented to ensure that such servers have consistent information. The Scalable Reliable Multicast protocol (SRM), discussed in section 6.4.1, is another example of a receiver-initiated NACK-avoidance protocol.

6.2.3 Tree-Based Protocols

Tree-based protocols divide the receiver set into local groups, distributing retransmission responsibility over an acknowledgment (ACK) tree structure. A simple illustration of a tree-based protocol is presented in Figure 6–3.

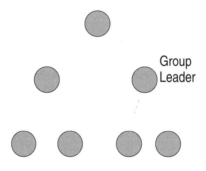

Figure 6–3 Tree-Based Protocols

Each local group consists of a group leader and a finite number of children. This number is independent of the size of the total receiver set and depends on the capacity of group leaders. An *ACK tree* is a structure of receivers (i.e., the hosts participating in a reliable multicast group) and is different from the underlying multicast routing tree(s) used by routers to multicast packets to the hosts.

Acknowledgments from children in a local group, including the source's own group, are sent only to the group leader. A child in a group sends its acknowledgment to its parent as soon as it receives a correct packet, not when all its own children (if any) have sent their acknowledgments. Clearly, these acknowledgments differ from ACKs or NACKs used in sender- and receiver-initiated protocols, and we refer to them as local ACKs. The use of local ACKs is very important for throughput; if the source had to wait for ACKs from the leaves to propagate all the way to the source, the pacing of the source would have to be based on the slowest path in the ACK tree.

Tree-based protocols delegate to leaders of local groups the decision of when to delete packets from memory, which is conditional upon receipt of local ACKs from the children in the group. Local ACKs are sent up subtrees composed of up to three types of nodes: a source node, leaf nodes, and group leaders. The source node is the originator of all new packets, which are multicast to all the receivers, using a multicast routing tree built by an underlying multicast routing protocol. The source and each group leader have a maximum number of children from which to process local ACKs and to which to send retransmissions. Leaf nodes are at the bottom of the ACK tree and are not responsible for any children; they play the same role as receivers in the sender-initiated protocol, except that they send their local ACKs only to their group leaders instead of sending ACKs to the source node. Group leaders send local ACKs to their own group leaders one step higher in the ACK tree and collect local ACKs from the children in their group, retransmitting if necessary.

The source and group leaders can control the retransmission time-outs, or such time-outs can be controlled by the children of the source and group leaders. In the first case, when the source sends a packet, it sets a timer, and each group leader sets a timer as it becomes aware of a new packet. If there is a time-out before all local ACKs have been received by the source or a group leader, the packet is assumed to be lost and is retransmitted by the source or group leader to its children. The retransmission strategy used is typically selective repeat, so that once a packet is received correctly, it is never rebroadcast to the group again.

It is also possible to use NACK-avoidance mechanisms within local groups to increase throughput in tree-based protocols. We refer to a tree-based protocol that uses NACK-avoidance and periodic polling [20] in the local groups as a tree-NAPP protocol. In this class of protocols, local NACKs are multicast within the local group to request retransmissions, and local ACKs are sent periodically by each child to confirm the reception of large partitions of data.

Most implementations of tree-based protocols assume that the source and group leaders indefinitely maintain the data transmitted in a session. However, tree-based protocols can be made to work correctly without requiring sources and group leaders to store all data sent in a session [13]. Section 6.5 discusses how this can be done.

Tree-based protocols eliminate the ACK implosion problem, free the source from having to know the receiver set, and operate solely on messages exchanged in local groups (between a node and its children in the ACK tree).

6.2.4   Ring-Based Protocols

Token-Ring-based protocols for reliable multicast were originally developed to provide support for applications that require an atomic and total ordering of transmissions at all receivers. One of the first proposals for reliable multicasting is the Token Ring protocol (TRP) [3]; its aim was to combine the throughput advantages of NACKs with the reliability of ACKs. The Reliable Multicast Protocol (RMP) [24] discussed an updated WAN version of TRP. Although multiple rings are used in a naming hierarchy, the same class of protocol is used for the actual rings.

We base our description of generic ring-based protocols on the LAN protocol TRP and the WAN protocol RMP. A simple illustration of a ring-based protocol is presented in Figure 6–4.

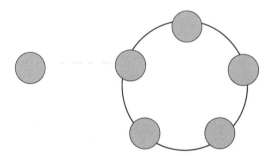

Figure 6–4 Ring-Based Protocols

The basic premise is to have only one token site responsible for ACK'ing packets back to the source. The source times out and retransmits packets if it does not receive an ACK from the token site within a time-out period. The ACK also serves to timestamp packets, so that all receiver nodes have a global ordering of the packets for delivery to the application layer. The protocol does not allow receivers to deliver packets until the token site has multicast its ACK. Receivers send NACKs to the token site for selective repeat of lost packets that were originally multicast from the source.

The ACK sent back to the source also serves as a token-passing mechanism. If no transmissions from the source are available to piggyback the token, then a separate unicast message is sent. The token is not passed to the next member of the ring of receivers until the new site has correctly received all packets that the former site has received. Once the token is passed, a site may clear packets from memory; accordingly, the final deletion of packets from the collective memory of the receiver set is decided by the token site and is conditional on passing the token. The source will only delete packets when an ACK/token is received. Note that both TRP and RMP specify that retransmissions are sent unicast from the token site.

6.3 Maximum Throughput of Reliable Protocols

Our discussion of maximum throughput that each of the generic reliable multi-cast protocols introduced in the previous section can achieve is based on the model first introduced by Pingali, Kurose and Towsley [18] [19]. Rather than focusing on communication bandwidth requirements, this model focuses on the processing requirements of generic reliable multicast protocols in a reliable multicast session that is already initialized and that experiences no failures of receivers or the source. Accordingly, the maximum throughput of a generic protocol is a function of the per-packet processing rate at the sender and receivers, and the analysis focuses on obtaining the processing times per packet at a given node.

We assume a single sender multicasting to R identical receivers. The probability of packet loss is p for any receivers. A maximum of B children per group leader or source is used for tree-based protocols. Because a single source is considered, a single ACK tree rooted at the source is assumed in the analysis of tree-based protocols. A selective repeat retransmission strategy is assumed in all the protocol classes since it is well known to be the retransmission strategy with the highest throughput. Because our comparison focuses on maximum attainable throughput of protocol classes, we will assume the best possible behavior of each protocol class. Accordingly, we assume that

- The timers required for NACK-avoidance in RINA and tree-NAPP protocols work perfectly, i.e., a single NACK needs to be sent for any lost packet and without incurring any control overhead.
- The token-passing scheme used in ring-based protocols incurs no overhead.

We make two additional assumptions for all protocol classes: (a) no ACKs or NACKs are ever lost and (b) all loss events at any node in the multicast of a packet are mutually independent. The first assumption is needed for NACK-avoidance mechanisms to work perfectly. This assumption favors protocol classes based on NACK-avoidance over their counterpart protocol classes based on ACKs because NACKs are multicast whereas ACKs are unicast. The second assumption is overly conservative and negatively impacts all protocol classes; it corresponds to the case in which there is no correlation of packet losses in the Internet, which is less accurate as the density of receivers increases with respect to the underlying multicast routing tree(s) used for packet and NACK dissemination.

Table 6–1 summarizes the bounds on maximum throughput for all the classes of reliable multicast protocols we consider. For the case of ring-based and tree-based protocol classes, the maximum throughput is obtained by analyzing the processing cost at each type of node participating in the protocol (e.g., source, group leader, or leaf nodes for tree-based protocols) and then obtaining the minimum of the maximum throughput obtained for each node type. Details on the derivation of the results listed in Table 6–1 are presented in [18][19][23][12][13].

Protocol	Processor requirements	p as a constant
Sender-initiated [18][19]	$O\left(R\left(1+\dfrac{p\ln R}{1-p}\right)\right)$	$O(R\ln R)$
Receiver-initiated with NAK-avoidance (RINA) [18][19]	$O\left(1+\dfrac{p\ln R}{1-p}\right)$	$O(\ln R)$
Ring-based [12]	$O\left(1+\dfrac{p(R-1)}{1-p}\right)$	$O(R)$
Tree-based [12]	$O(B(1-p)+pB\ln B)$	$O(1)$
Tree-based with local NAK-avoidance (Tree-NAPP) [13]	$O\left(1+\dfrac{1-p+p\ln B+p^2(1-4p)}{1-p}\right)$	$O(1)$

Table 6–1   Analytical bounds on throughput for R receivers and probability of packet loss p.

The results in Table 6–1 show that tree-based, tree-NAPP, and RINA protocols are the most attractive approaches.

Even as the probability of packet loss goes to zero, the throughput of the sender-initiated protocol is inversely dependent on R, the size of the receiver set, because an ACK must be sent by every receiver to the source once a transmission is correctly received. In contrast, as p goes to zero, the throughput of receiver-initiated protocols becomes independent of the number of receivers. Notice, however, that the throughput of a receiver-initiated protocol is inversely dependent on R, the number of receivers, or on $\ln R$, when the probability of error is not negligible. For the case of ring-based protocols, the maximum throughput is inversely dependent on the number of receivers when p is not zero. Of all the classes of reliable multicast protocols that have been proposed, tree-based and tree-NAPP protocols are the only classes whose maximum attainable throughput is completely independent of the number of receivers.

A better way to visualize the maximum attainable throughput of protocol classes is by plotting the number of supportable receivers by each of the different classes, relative to processor speed requirements. This number is obtained by normalizing all classes to a baseline processor, as described in [13][23]. The baseline uses the sender-initiated protocol and can support exactly one receiver. Given the processor speed required to support a single receiver in the sender-initiated protocol, the curves shown in Figure 6–5 indicate the increase in processor speed needed to support different numbers of receivers in each protocol class. The curves shown in Figure 6–5 assume 10 children per local group in tree-based protocols. It is clear from the figure that RINA protocols, tree-based protocols, and tree-NAPP protocols are the better approaches.

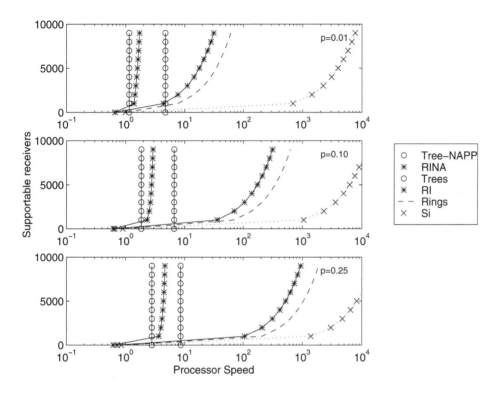

Figure 6–5 Number of Supportable Receivers for Each Protocol
The probability of packet loss is 1%, 10%, and 25% respectively. The branching factor for trees is set at 10.

6.4 Protocol Implementations

From the previous section, it appears that RINA protocols and tree-based protocols are the most attractive choices from the standpoint of maximum throughput that they can achieve. In fact, receiver-initiated protocols and tree-based protocols are the two types of protocols being considered for reliable multicasting in the Internet today. We describe three such protocols in this section.

6.4.1 Scalable Reliable Multicast (SRM)

The generic RINA protocol we have just described constitutes the basis for the operation of the scalable reliable multicasting (SRM) algorithm [8]. SRM has been embedded into an internet collaborative whiteboard application called *wb*. SRM sets timers based on low-rate, periodic "session-messages" multicast by every member of the group. The messages specify a timestamp used by the receivers to estimate the delay from the source, and the highest sequence number generated by the node as a source. The average bandwidth consumed by session messages is kept small (e.g., by keeping the frequency of session messages low). SRM's implementation requires that every node stores all packets or that the application layer stores all relevant data.

NACKs from receivers are used to advance the *cw*, which is controlled by the receivers, and the sequence number in each multicast session message is used to "poll" the receiver set, i.e., to ensure that each receiver is aware of missing packets. Although session messages implement a "periodic polling" function [20], they cannot be used to delete packets, as in a sender-initiated protocol, because a sender specifies its highest sequence number as a source, rather than the highest sequence number heard from other sources.

In practice, the persistence of session messages forces the source to process the same number of messages that would be needed for the source to know the receiver set over time (one periodic message from every receiver). Accordingly, as defined, SRM does not scale, because it defeats one of the goals of the receiver-initiated paradigm, i.e., to keep the receiver set anonymous from the source for scaling purposes.

Other issues limit the use of SRM and other RINA protocols for reliable multicasting. First, SRM requires that data needed for retransmission be rebuilt from the application. SRM's approach is reasonable for applications in which the immediate state of the data is exclusively desired, which is the case of a distributed whiteboard. However, the approach would not apply for multimedia applications that have no current state, but have only a stream of transition states.

Second, NACKs and retransmissions must be multicast to the entire multicast group to allow suppression of NACKs. The basic NACK-avoidance algorithm requires that timers be set based on updates multicast by every node, which implies that each node must do an increasing amount of work as the number of nodes increases! Furthermore, nodes that are on congested links, LANs, or regions might constantly bother the rest of the multicast group by multicasting NACKs. Accordingly, the NACK-avoidance scheme is well-suited for a limited scope, such as a LAN, or a small number of Internet nodes (as it is used in tree-NAPP protocols, described in the next section).

RINA protocols can achieve maximum throughputs that are fairly independent of the number of receivers when timers are set exactly, and mechanisms can and have been designed to suppress NACKs as much as possible. However, NACK-avoidance mechanisms depend on the setting of timers at the receivers, the accuracy of which is critical and which can be easily disrupted by packet and NACK losses, link and path asymmetries, fluctuations in path delays, and changes in the membership of the group or the underlying multicast routing tree(s). If error recovery in a RINA protocol depends solely on time-outs at the receivers, end-to-end delays can become arbitrarily large.

Approaches to limit the scope of NACKs and retransmissions have been proposed for SRM [8]; however, these approaches rely on session messages multicast to all group members. Limiting the scope of session messages requires their aggregation into local group messages, so that the source does not receive as many of such messages. This approach, however, leads to the establishment of some receiver structures, which may well be an ACK tree.

6.4.2 Tree-Based Protocols

The first application of tree-based protocols to reliable multicasting over an internet was reported by Paul et al. [16], who compare three basic schemes for reliable point-to-multipoint multicasting using hierarchical structures. Their results have been fully developed as the reliable multicast transport protocol (RMTP) [14][17]. RMTP has been implemented on several platforms and has been used successfully in AT&T's "call detail data distribution network."

The tree-based multicast transport protocol (TMTP) [25] is the only specification of a tree-NAPP protocol to date. Messages sent for the setting of timers needed for NACK-avoidance are limited to the local group, which is scalable.

RMTP, TMTP, and existing implementations of tree-based protocols also present scaling problems. RMTP and TMTP require the source and group leaders to maintain indefinitely all data sent in a session, and require an ACK tree for each source of the reliable multicast group. As we have pointed out, not all applications can be asked rebuild the data needed for retransmission, and maintaining an ACK tree per source becomes onerous when there are many sources per group. Details of RMTP and TMTP are provided below.

Reliable Multicast Transport Protocol (RMTP)

RMTP provides a sequenced, reliable bulk transfer service between one source and many receivers. Like all tree-based protocols, RMTP designates three types of hosts: the *source, designated receivers* (group leaders), and *receivers* (leaves). All hosts are placed in an ACK tree structure constructed on the basis of the hop distance between hosts.

RMTP is based only on a congestion window; deletion of packets must be handled by the application. The congestion window is advanced by periodic ACKs sent to group leaders from children of each local group. The ACKs represent each receiver's own status rather than the aggregate of all its children and allow the group leaders to determine which packets need to be retransmitted. RMTP uses a *selective repeat* retransmission strategy. ACKs contain the sequence number of the last packet received correctly, as well as a bit vector representing the status of the remaining set of packets in the current window. If the number of children missing a particular packet is above a set threshold, then the retransmission is multicast to the local group; otherwise, it is unicast to the individual receiver.

ACK tree and local group construction in RMTP on the Internet is based on the hop counts between the hosts in the session. Periodically, each group leader and the source multicast a SEND_ACK_TOME packet to all the receivers. Each SEND_ACK_TOME packet from each group leader contains the same initial time-to-live (TTL) value in its IP-header, which is decremented at each router as the packet is delivered to each host. Receivers may collect several SEND_ACK_TOME packets but pick only a single group leader that is the source of the SEND_ACK_TOME with the largest remaining TTL value, since this is the closest group leader.

RMTP uses a window-based flow control mechanism. The source sends only a window of W_s packets during a time period T_{send}, limiting the maximum transmission rate to W_s*packet_size/T_{send}. The source does not advance the congestion window based on its current set of children, but rather based on the set of group leaders that have sent status messages within some interval. The window advance is based on all the receivers that have ACK'ed in that interval. Note that when the window is advanced, the packets are cached rather than deleted. This approach enables RMTP to service late-joining receivers easily.

RMTP depends on the services of a *session manager* to manage initial connection parameters of a session. When receivers agree to the connection parameters, they start the session in an *unconnected* state. Once the first data packets are received from the sender, receivers change to a *connected* state in which ACKs are periodically sent to the group leader. The group leader considers the connection live for as long as it receives periodic ACKs.

The source does not keep track of the receiver set, nor do receivers inform that source when they leave the session. When the sender transmits the final packet, a timer T_{dally} is started, which is set to at least twice the lifetime of a packet on the Internet. Similarly, local group leaders also keep a timer when they receive the final packet. Receivers simply leave the session, with no notice to other hosts, when the last packet has been received correctly. When each group leader's T_{dally} timer expires, including the source, the session ends. Any ACK received at the source or group leaders resets the timer, allowing receivers to continue the session. Note that temporary network partitions and situations that may cause a receiver to be temporarily disconnected from the group until the session ends are assumed to be handled by the session manager. However, since packets are not deleted at the source, hosts that can reconnect before the session ends can request any missing packets.

Tree-Based Multicast Transport Protocol (TMTP)

TMTP constructs an ACK tree on the basis of *expanding ring searches* (ERS). In an ERS, apacket is sent to the multicast group using a limited scope based on a small TTL value that prevents the packet from reaching every receiver in the session. If a response is not heard, then the packet is retransmitted using a larger TTL value in order to increase the scope of the multicast. Similarly to RMTP, TMTP has each receiver send periodic ACKs to local group leaders. However, TMTP also calls for the use of a NACK avoidance scheme within each local group.

Tree construction starts with each receiver multicasting a SEARCH_FOR_PARENT message with a small time-to-live (TTL) value in the IP packet header. In the Internet, the TTL field of every packet is decremented as it traverses each router. Packets are dropped when the TTL field reaches 0. If no hosts reply with a WILLING_TO_BE_PARENT message within a fixed time-out, a new SEARCH_FOR_PARENT message is sent with a larger TTL value, in other words, ERS is used. This process is repeated until a response is heard. The hope is, that by expanding the scope of the messages, the closest nodes willrespond first and the amount of network traffic will be reduced. Group leaders in the ACK tree are selected on the basis of the closest hop count.

Groups are not segregated into separate multicast addresses. Each group leader maintains a table of the distance to each child, in hop counts. This maximum distance, termed the *maximum radius*, determines the scope (i.e., the TTL value) of retransmissions.

TMTP uses window-based flow control techniques with a predefined static maximum transmission rate in order to avoid bursty traffic. The focus on TMTP flow control is to allow an abundant amount of time for ACKs to be received at the source and group leaders to avoid sending multicast retransmissions. Even though the retransmissions are of a limited scope, they still may waste bandwidth and be a source of congestion on a large area of the network topology. Accordingly, the size of each window is calculated as a multiple of a timer T_{ack}, which is the longest round-trip time from the group leader to its children. In other words, the size of the window is set by the number of packets that can be sent during the period $T_{retrans} = n \times T_{ack}$, where n is an integer greater than 1 and recommended to be 3. Packets sent during each T_{ack} interval are not retransmitted until the interval ends.

Since the TMTP window will not advance without an ACK from each child, children will cause their group leader's windows to fill up if they do not send ACKs fast enough. In turn, the group leader will be unable to send ACKs to its parent, causing *backpressure* to flow toward the source.

6.5 Scaling and Efficiency Issues

In this section, we describe additional mechanisms for reliable multicast protocols that address some of the limitations of existing protocol implementations for reliable multicasting.

The mechanisms we summarize are based on the Lorax protocol [13], which we use as an example of the issues that must be considered to make tree-based protocols more scalable and efficient.

6.5.1 Deallocating Memory

Today's implementations of RINA and tree-based protocols require sources (and group leaders, in the case of tree-based protocols) to maintain all data sent in a session indefinitely, and this approach need not be acceptable depending on application. Today's ring-based protocols do allow the source to delete data from memory in a finite time but are not scalable with the number of receivers.

The problem with allowing the source to delete data from memory after a finite time is that the source should not be allowed to do that until it is certain that all receivers participating in the multicast group have received the data correctly even when receivers fail and recover.

Consider the case of tree-based protocols first. After node failures, applications could either (a) terminate the session, (b) continue the session without providing a reliable service to nodes while (and in some cases after) they are temporarily disconnected, or (c) allow nodes to rejoin and catch up with the session. For the first two cases, changes in the ACK tree do not create a problem, and the cw and mw can both advance together at the source and each group leader. The third case is a problem, however, and requires the basic tree-based protocol class to be modified slightly.

Figure 6–6 illustrates the problem with an example.

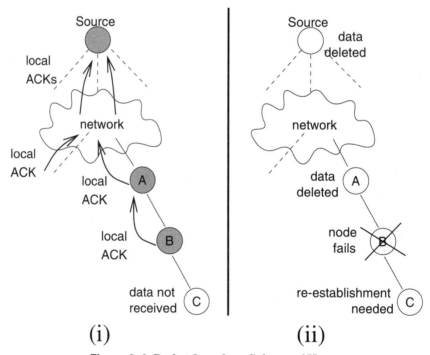

Figure 6–6 Packet Loss for a Subtree of Hosts
Loss can occur when aggregate ACKs are not used and there is a failure at a group leader.

Packets are multicast from the source to the receiver set; nodes that have received the data correctly are shaded. Packets are acknowledged as they are received, rather than waiting for the acknowledgments from children. Node A and all other nodes that have received local ACKs from all their children delete packets; however, B fails before it is able to confirm that all its children have correctly received the data. If we assume that at least one child C has not received the data, then there is no node with which to reestablish contact that will definitely have a copy of the data. One solution to this problem is to buffer the entire session in a secondary store, as is done in RMTP and SRM. However, this solution can become unscalable.

Ideally, we would like to keep data in a finite secondary store only until all current receivers have correctly received the data, without having any node keeping track of all the receivers. Fortunately, deadlocks due to receiver failures, reconfigurations of the ACK tree, or changes to the underlying multicast routing tree can be easily avoided in tree-based protocols by introducing aggregate ACKs that propagate from the receivers up the tree to the source. The aggregate ACK sent from a node to its parent in the tree consists of its own ACK and the aggregated ACKs from all its children. Just as in the generic tree-based protocol, correctly received data packets are acknowledged by local ACKs. However, packets are not deleted at this point; instead, they are kept in a secondary store or partition of memory. When a parent of a leaf node confirms that all its children have correctly received the data, it deletes the data from secondary store and sends an aggregate ACK to its own parent. Group leaders carry out the same procedure. In terms of our taxonomy, aggregate ACKs are used to move the mw, and local ACKs (and local NACKs) to move the cw.

Two additional mechanisms are needed together with aggregate ACKs to ensure that a disconnected node or subtree is never allowed to rejoin the ACK tree after the source has erased data from memory that the rejoining node or subtree never received. First, a node that perceives one of its children as disconnected assumes the reception of any pending aggregate ACK from that child and sets a topology-change notification flag in its own aggregate ACKs. The setting of the flag is preserved as the aggregate ACK travels back to the source. The flag instructs the source to wait for an even longer period of time before erasing the associated data from memory after receiving all the aggregate ACKs from its children. Second, an orphan node is given a finite amount of time to reconnect to the ACK tree. This time is much shorter than the time set in a "connect timer" at the source. Once an orphan node times out, it cannot join the session and catch up without application-level support. Details of how these mechanisms can be implemented are given in [13].

Consider now the case of RINA protocols. The same mechanisms based on sending aggregate ACKs over an ACK tree to ensure safe deletion of data packets at the source can be used [13]! A ring structure would not be as attractive because of the delays involved in traversing a ring of many receivers. The frequency with which aggregate ACKs are transmitted need not be any different than the frequency with which session messages are transmitted in SRM, for example.

6.5.2 Common ACK Trees

For a concurrent multicast session with many sources, it is not reasonable to manage a separate ACK tree for every source. To remedy this, Levine, Lavo, and Garcia-Luna-Aceves [13] have proposed the dissemination of all ACKs in a multicast group along a single, shared ACK tree of the concurrent multicast session. This scheme is equivalent to the use of a single ring in a ring-based protocol.

The mechanisms used to build and maintain a shared ACK tree itself are essentially the same as those proposed and implemented for existing tree-based protocols. Details of these mechanisms can be found in [13]. The main difference between previous tree-based protocols and protocols based on shared ACK trees lies in the handling of ACKs along the ACK tree. When an ACK tree is created for a single-source multicast session with the source as the root of the tree, the routing of aggregate ACKs to the appropriate group leader towards the source is simple: each node sends local ACKs to its designated parent. However, the situation is more complicated for shared ACK trees. The introduction of multiple sources clashes with the inherent anonymity of the tree: receivers in the ACK tree lack knowledge of where each source is located, knowledge that can be used to route ACKs to the appropriate group leader towards a particular source.

Routing of ACKs in an anonymous ACK tree can be done efficiently by implicit routing, with which each node is labeled based on its position in the tree relative to a tree root. Because a packet contains the source's label, a receiver can determine if the ACK for a packet it receives should be routed to its parent or one of its children in the ACK tree by comparing its own label with the label of the source specified in the packet. The labeling scheme illustrated in Figure 6–7 was introduced for the Lorax protocol [13]. The scheme involves only two nodes for a completely new addition (the added node and its parent), and deletions require relabeling of the subtree of the deleted node when patched back into the ACK tree. According to this labeling scheme, a receiver routes ACKs to its parent if the label of the packet's source does not contain the receiver's label as a prefix. Otherwise, the next character in the label determines the receiver's child to which the ACK should be routed to reach the packet's source.

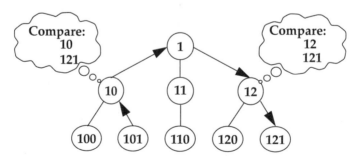

Figure 6–7 An Example Labeling Scheme
Members 10 and 12 decide how to route ACKs to
source 121.

In the example shown in Figure 6–7, arrowheads indicate the direction that ACKs to node 121 would follow from a few nodes. Shared ACK trees can be used in the background of RINA protocols in order to allow the sources to safely delete data from memory.

6.5.3 Efficient Ack Tree Construction

The main problem in constructing an ACK tree consists of reducing the amount of traffic needed for sources and receivers to join the tree. Lorax [13] utilizes a combination of root-based and off-tree schemes to grow an ACK tree. These schemes are based on an expanded ring search (ERS) over the underlying multicast routing tree(s) and include mechanisms intended to limit the cost of each ERS.

The ACK tree is grown from a single root node, using either the source multicast routing tree of the root node or the common multicast routing tree of the multicast group. The root node may be selected before the session starts and advertised together with the multicast address or may be selected by means of an election algorithm when the session begins.

After joining the IP multicast address, all nodes are considered *off-tree* except for the root node of the ACK tree. The root immediately begins multicasting invitation-to-join messages (INVs) with a time-to-live (TTL) value of zero in the IP header, and sets a timer. An off-tree node that hears an INV message unicasts a request-to-be-adopted message (REQ) back to the inviting node. If an inviting node does not hear a REQ before the timer expires, it multicasts a new INV with a larger TTL value and resets the timer with a longer time-out. When a REQ is received correctly at the inviting node, a bind message (BIND) is sent in response to the new child confirming the adoption. Once the new child receives the BIND, it becomes an *on-tree* node and starts the same process again by multicasting an INV. This process stops at any on-tree node (i.e., a node that is "growing the ACK tree") when the node has several children or the TTL field of its INV reaches a maximum value.

The maximum TTL of an INV is much smaller than the TTL of data packets or the TTL needed to cover the entire underlying multicast routing tree. This root-based strategy to create the ACK tree avoids excessive traffic over the multicast routing tree. Lorax also includes an off-tree scheme for off-tree nodes to reach the ACK tree to permit any receivers to request reliable transmissions, even if the invitations to join the reliable multicast group fall short of reaching all interesting receivers;, details are presented in [13].

6.5.4 Tree Maintenance

The ACK tree can change because of additions of new group members, departures or failures of members, or network partitions. The analysis in section 6.3 (also see [12]) assumed the same number of children for each group leader in the ACK tree. In practice, the number of children are chosen independently for each group leader because some machines have more processing power available than do others. For reasonable performance, a group leader should not be limited to having too few children, and it should not have more children than the number it can support efficiently.

We need an algorithm to keep the number of children at or below a given value at each group leader; we refer to this process as *fissioning*. A simple heuristic is to simply disconnect extra children and let them ask nodes in other local groups for adoption; however, this solution may create unscalable amounts of work at some nodes.

The fission algorithm used in Lorax [13] requires that parent nodes keep track of how many additional nodes its current children may take. This information is easily included periodically in local ACKs or as part of a NAPP algorithm. The parent node sends an adopt message (ADOPT) to the child with the most free space. The ADOPT forces the adopting node to multicast to all the nodes in its local group in the ACK tree a request-for-children message (RFC). The nodes that respond first are currently closer and become children of the adopting node, and the fission is completed. From here, the labeling algorithm must be run on the new subtrees of the adopting nodes.

6.6 Summary

We have compared known classes of reliable multicast protocols. It is already known that sender-initiated protocols are not scalable at all since the source must account for every receiver listening. Receiver-initiated protocols are more scalable, especially when NACK-suppression schemes are used to avoid overloading the source with retransmission requests. However, this protocol class can be used efficiently only with application-layer support and only for some applications because of the unbounded-memory requirement. Current ring-based protocols were designed for atomic and total ordering of packets but are not scalable.

Our analysis shows that trees are a good approach to the scalability problem for reliable multicasting. Of the protocol classes that have been proposed, only tree-based and tree-NAPP classes have a maximum throughput that is constant with respect to the number of receivers, even when the probability of packet loss is not negligible.

Tree-based protocols delegate responsibility for retransmission to receivers and employ techniques applicable to either sender- or receiver-initiated protocols within local groups (i.e., a node and its children in the tree) of the ACK tree only. Therefore, any mechanism that can be used in a receiver-initiated protocol can be adopted in a tree-based protocol, with the added benefit that the throughput and number of supportable receivers is completely independent of the size of the receiver set, regardless of the likelihood with which packets are received correctly at the receivers.

Future work on reliable multicasting for the Internet is likely to include receiver structures. ACK trees appear ideally suited for the job, but more research is needed to determine what constitutes the most scalable and efficient approach for structuring receivers. Research questions that must be addressed, in addition to those discussed in section 6.5, are security of reliable multicast groups, flow control of multicast sessions, and interoperation of multiple reliable multicast protocols; t is unlikely that a single solution will satisfy all types of networking environments and user communities. Lastly, we point out that throughout this chapter we have assumed that the protocols used for end-to-end reliable multicast operate independently from the protocols used for multicast routing. However, a closer interaction among these protocols can only simplify their tasks and improve their performance. The development of IPv6 [9] and the early stages of deployment of both multicast routing protocols and end-to-end reliable multicast protocols offer a good opportunity for new approaches that foster a closer collaboration across protocol layers.

Acknowledgments

This work was supported by the Defense Advanced Research Projects Agency (DARPA) under Grant F19628-96-C-0038.

References

[1] A. Ballardie, P. Francis, and J. Crowcroft, "Core Based Trees (CBT): An Architecture for Scalable Inter-Domain Multicast Routing," *Proc. ACM SIGCOMM* 1993:85–95.

[2] D. Bertsekas and R. Gallager, *Data Networks*. 2nd ed. Englewood Cliffs, NJ: Prentice Hall,1992.

[3] J. M. Chang and N.F. Maxemchuk, "Reliable Broadcast Protocols," *ACM Transactions on Computer Systems* 2(3):251–273, 1984.

[4] D. D. Clark, M. L. Lambert, and L. Zhang, "NETBLT: A High Throughput Transport Protocol," *Proc. ACM SIGCOMM* 1993:353–359.

[5] S. Deering, *Multicast Routing in a Datagram Internetwork*, Ph.D. thesis, Stanford University, Palo Alto, California, December 1991.

[6] S. Deering and D. Cheriton, "Multicast Routing in Datagram Inter-networks and Extended LANS," *ACM Transactions on Computer Systems* 8(2):85–110, 1990.

[7] Deering, S., D. Estrin, D. Farinacci, V. Jacobson, C.-G. Liu, and L. Wei, "Architecture for Wide-Area Multicast Routing," *Proc. ACM SIGCOMM* 1994:126–135.

[8] S. Floyd, S., V. Jacobson, S. McCanne, L. Zhang, and C.-G Liu, "A Reliable Multicast Framework for Light-Weight Sessions and Application Level Framing," *Proc. ACM SIGCOMM* 1995: 342–356.

[9]   R. Hinden and S. Deering, *IP Version 6 Addressing Architecture*, RFC 1884, December 1995.

[10]   H. Holbrook, S. Singhal, and D. Cheriton, "Log-based Receiver-Reliable Multicast for Distributed Interactive Simulation," *Proc. ACM SIGCOMM* 1995:328–341.

[11]   J.B. Postel, editor, *Transmission Control Protocol*, RFC 793, September 1981.

[12]   B. Levine and J. J. Garcia-Luna-Aceves, "A Comparison of Known Classes of Reliable Multicast Protocols," *Proc. IEEE International Conference on Network Protocols* 1996:112–121.

[13]   B. Levine, D. Lavo, and J. J. Garcia-Luna-Aceves, "The Case for Reliable Concurrent Multicasting Using Shared ACK Trees," *Proc. ACM Multimedia* 1996:365–376.

[14]   J. Lin and S. Paul, "RMTP: A Reliable Multicast Transport Protocol," *Proc. IEEE Infocom* 1996:1414–1425.

[15]   M. Parsa and J. J. Garcia-Luna-Aceves, "A Protocol for Scalable Loop-Free Multicast Routing," *IEEE Journal on Selected Areas in Communications* 15(3):316–331, April 1997.

[16]   S. Paul, K. Sabnani, and B. Kristol, "Multicast Transport Protocols for High Speed Networks," *International Conference on Network Protocols*, 1994:4–14.

[17]   S. Paul, K. Sabnani, J. Lin, and S. Bhattacharyya, "Reliable Multicast Transport Protocol (RMTP)," *IEEE Journal on Selected Areas in Communications* 15(3):407–421, April 1997.

[18]   S. Pingali, *Protocol and Real-Time Scheduling Issues For Multimedia Applications*, Ph.D. thesis, Jain University of Massachusetts, Amherst, September 1994.

[19]   S. Pingali, D. Towsley, and J. Kurose, "A Comparison of Sender-Initiated and Receiver-Initiated Reliable Multicast Protocols," *Performance Evaluation Review* 22:221–230, May 1994.

[20]   S. Ramakrishnan and B.N. Jain, "A Negative Acknowledgment with Periodic Polling Protocol for Multicast over LAN," *Proc. IEEE Infocom* 1987:502–511.

[21]   C. Shields and J. J. Garcia-Luna-Aceves, "The Ordered Core-Based Tree Protocol," *Proc. IEEE Infocom*, April 1997.

[22]   T. Strayer, B. Dempsey, and A. Weaver, *XTP: The Xpress Transfer Protocol*. Menlo Park, CA: Addison-Wesley Publishing Company, 1992.

[23]   D. Towsley, J. Kurose, and S. Pingali, "A Comparison of Sender-Initiated and Receiver-Initiated Reliable Multicast Protocols," *IEEE Journal on Selected Areas in Communications* 19(3):398–420, April 1997.

[24]   B. Whetten, S. Kaplan, and T. Montgomery, "A High Performance, Totally Ordered Multicast Protocol," *Theory and Practice in Distributed Systems*, International Workshop, LNCS 938, 1994.

[25]   R. Yavatkar, J. Griffioen, and M. Sudan, "A Reliable Dissemination Protocol for Interactive Collaborative Applications," *Proc. ACM Multimedia* 1995:333–44.

Multimedia Applications in Networks

Torsten Braun

In this chapter, we describe new multimedia applications mainly developed to run over the Internet. Several tools for multiparty audio/video conferencing have been implemented during the last few years, based on the Application Level Framing concept (ALF). ALF is a new design concept for multimedia applications and communication systems. The chapter discusses the architectural concepts of the conferencing tools and solutions to provide QoS for heterogeneous receivers. Video servers and shared whiteboard and editing tools are other important applications upcoming in the Internet. Approaches for the integration of multicast and real-time data transfer into the WWW are also illustrated. Future "killer" applications might be interactive multimedia applications, e.g., for teleteaching and interactive multiuser games requiring both reliable multicast and real-time communication support. Early examples of those applications are presented.

7.1 Introduction

Several years ago, data, video, and audio were always transferred over separate networks. Telephone networks and private branch exchanges have been used for audio transfer. Data such as text or files has been transmitted over classical data networks, e.g., public X.25 networks or local area networks. The Internet was used by scientific and educational institutions only for data communication applications such as file transfer, electronic mail, terminal emulation, net news, etc. These data applications were and still are running over TCP/IP or UDP/IP Internet communication systems. The users were using terminals connected to a host computer or personal computers without any possibility of processing audio or video data.

Along with the evolving multimedia capabilities of personal computers and workstations (e.g., high-resolution color graphics and audio/video (A/V) equipment) computers have become very useful in processing multimedia data (graphics, video, audio, data, and text) and in exchanging the data over the Internet. Moreover, the traditional communication paradigm with two communicating users has changed, and the demand to support efficient communication among groups of users has increased significantly. The communication protocols being used in the Internet supported neither multicast communication nor real-time data transfer efficiently.

Multicast functionality was added to the IP protocol at the end of the previous decade[16][48]. IP hosts and routers were upgraded with IP multicast code and interconnected to the so-called Multicast Backbone (MBone). The most popular MBone applications have been audio/video and shared whiteboard conferencing applications. These multicast tools offered services that have not been possible with standard telephone technology, in particular, on-demand audio/video group communication. However, in addition to UDP/IP, these tools require functions such as synchronization support. This requirement led to the development of the Real-Time Transport Protocol (RTP). Multicast-capable UDP/IP communication stacks enhanced by RTP provide the basis for many Internet multimedia applications today.

7.2 Application-Level Framing

As explained in Chapter 5, TCP and UDP were initially developed for data transfer applications, but they do not provide any support for real-time data such as audio and video. TCP is even inappropriate for real-time data transfer because it provides a reliable and sequenced service to the applications. Packet losses are recovered by retransmissions and because of the sequenced delivery at the receiver, data delivery is delayed until all retransmissions of prior data have been received. This process introduces an intolerable delay in the case of packet loss.

UDP avoids that drawback, offering a very simple datagram-oriented but unreliable service. The simplicity makes UDP attractive for a variety of new applications. All functionality required on top of UDP can be integrated into the applications. This approach is very attractive for efficient communication system implementations. The concept of integrating special communication system functions has been proposed by the Application Level Framing (ALF) concept developed by Clark and Tenenhouse [13].

The integration of traditional transport protocol functionality such as reliable delivery requires that the application controls the packet size used in the network. Clark and Tenenhouse propose to use a single packet size for all functions performed in the communication stack [13]. Experiments have proved that the ALF concept improves the communication systems performance and allows more advanced techniques for the efficient implementation of communication systems. To support interoperability among the applications, ALF protocol frameworks define mainly the protocol data unit formats. Algorithms such as those required for flow control or retransmissions are not defined mandatorily by the protocol framework but are designed as depending on application requirements. Two popular protocol frameworks in line with the ALF concept are the Real-Time Transport Protocol (RTP) and Scalable Reliable Multicast (SRM).

Real-time applications such as adaptive audio or video conferencing tools are often based on RTP so that they can work in loaded networks as long as a minimal amount of bandwidth is available. Real-time applications make use of RTP's support of intra- and interstream synchronization and encoding detection. RTP is often integrated into the application software rather than being implemented as a separate layer. According to the ALF principle, the semantics of several RTP header fields are application dependent. Several profile documents specify the use of the RTP header fields for different applications. For example, the marker bit of the RTP header defines the start of a talk spurt in an audio packet and the end of a video frame in a video packet.

Reliable multicast applications require functions to provide reliable communication in UDP/IP-based multicast scenarios. Chapter 6 provides an overview of reliable multicast transport protocols. One of these approaches is the SRM protocol framework [20], which has been implemented in the shared whiteboard conferencing tool, wb. SRM provides algorithms increasing the multicast reliability over unreliable UDP/IP communication systems. SRM has not been standardized yet, but there is a proposal to enhance the RTP data format in order to include the required information elements for retransmission requests and retransmission packets.

Proceeding from the ALF concept, a group of researchers at LBL (Lawrence Berkeley Laboratory) developed the Light-Weight Session (LWS) model as a generalization of IP multicast for collaborative applications such as A/V conferencing and shared whiteboards. The goal of the model is to enhance IP multicast by scalability, fault tolerance, and robustness. The LWS building blocks are IP multicast, timing recovery via receiver adaptation, and thin transport layers according to the ALF concept [29]. The Scalable Reliable Multicast (SRM) framework can be considered as an extension of the LWS principle to support reliable multicast data transfer.

7.3 Audio/Video Conferencing

During the last years, many multimedia applications have been developed for the Internet. The MBone tools are some of the most popular ones. Several video and audio conferencing tools allow video and audio conferences among two or more users.

A transmission of an IETF meeting in March 1992 was the first major MBone event. Since then, almost all of the IETF conferences have been multicasted over the MBone, using the tools vat, vic, and wb, which are described in the following sections. MBone participants can watch and listen to the IETF sessions and are able to ask questions interactively.

On November 18th, 1994, a 25-minute concert of the Rolling Stones was multicasted over the Internet. The Rolling Stones used this event, which was one of the first commercial ones, as a promotion for their pay-per-view offer one week later. Also, live videos from NASA space shuttle missions or events such as surgeries have been multicasted. Today, the most frequent application of these tools are transmissions of university lectures and teleseminars.

7.3.1 Session Directories

Since transmissions over the MBone are becoming more popular, MBone users often have the choice between simultaneous live transmissions. Furthermore, a conference requires the starting of several applications with specific parameters (e.g., IP multicast addresses and UDP port numbers) which users often do not understand. These problems are addressed by session directory applications such as sd and sdr [24]. The session directory tools behave like TV guides and are used to announce and start MBone conferences.

First, a conference must be generated by the configuration of several parameters such as destination IP multicast addresses for each medium, corresponding UDP port numbers, encoding schemes for audio/video and program names of other applications such as whiteboards.

Once created, the name of the conference is advertised by sd or sdr. The user selects a conference by a mouse double-click, and all required applications start automatically with the announced parameters. For audio and video applications, it is not required that all participants use exactly the same tools. A participant must have at least one tool for each stream supporting the announced audio/video formats. The tools supporting these formats are then selected by sd or sdr.

Figure 7–1 contains several sdr windows. The small window shows the basic sdr window announcing the currently offered sessions. In that example, only one conference is offered. The details of this conference are shown in the upper-left window. In the right window, a user is currently creating a second conference entry. Both conferences consist of PCM audio, H.261 video, the shared whiteboard tool wb, and the network text editor from University College of London (UCL).

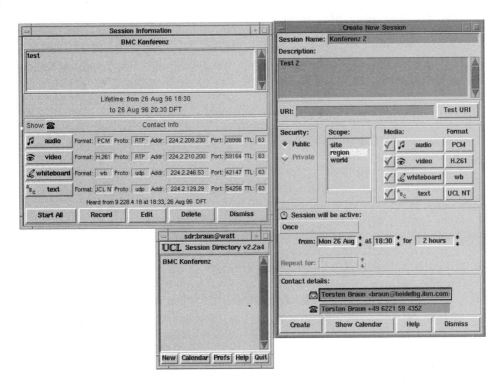

Figure 7–1 Session Directory (sdr)

7.3.2 Audio/Video Conferencing

The network video (nv) tool [21] developed at Xerox PARC and the INRIA video conferencing system (IVS) [46] are the earliest multicast-capable Internet video conferencing tools. IVS also supports audio transfer. The video conferencing tool (vic) [35] was implemented at LBL and based on the experiences gained with NV and IVS. Vic is, today, the most widely used video conferencing tool on the MBone. The Network Voice Terminal (Nevot) [41] from University of Massachusetts and the visual audio tool (vat) [28] from LBL are probably the most well-known audio tools.

Most of these tools are based on RTP and use a standardized packet format to encapsulate different audio and video encoding schemes and transmit them over UDP and multicast IP. The different tools significantly influenced the RTP standardization process [42], with the result that most of the tools can interoperate with each other. RTP is also the basis for many audio/video conferencing products commercially available today.

Vic: The Video Conferencing Tool

The video conferencing tool vic was developed at LBL, based on experiences gained with its predecessor tools nv and ivs. Vic can run over different network layers such as UDP/IP, the Tenet protocol stack [5], or AAL5/ATM. Whereas ivs and nv support video compression and decompression in software only, vic has been designed to also take advantage of hardware compression and decompression. Another nice feature of vic is its modular software architecture that enables vic to extend or modify its user interface.

A demultiplexer receives arrived packets from the network and implements the bulk of RTP processing (Figure 7–2). From there, the packet is demultiplexed to an appropriate decoding component. Software and hardware decoding, multiple compression formats, and different output devices such as external video output and X Window System™ devices are supported. A similar architectural decomposition was chosen for the capture/compression path.

To provide confidential communication, vic implements end-to-end encryption according to the RTP specification. Encryption is the last processing step in the sending path, and decryption is the first step in the reception path. By default, the Data Encryption Standard (DES) in cipher chaining mode is taken.

Vic is currently being used in many teleteaching projects such as the Heidelberg-Mannheim project. One of the first projects of this kind was the distribution of course lectures over the U.C. Berkeley campus network in 1994. In the same year, a live surgery performed at U.C. San Francisco Medical School was transmitted to a European medical conference.

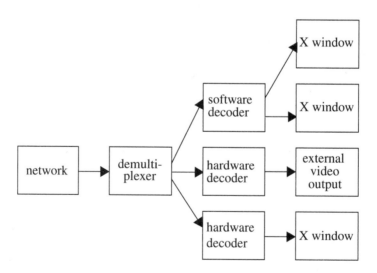

Figure 7–2 Structure of the vic Receiver

Vat: The Visual Audio Tool

Vat is a real-time Internet audio conferencing application. It can run point-to-point using standard unicast IP addresses, but it is primarily intended as a multiparty conferencing application. Vat is probably the most widely used Mbone audio tool and is available for most UNIX® platforms and Microsoft® Windows™. It enables the users to see who is actively speaking by highlighting the speaker's identity field. Identity information is exchanged by RTCP control packets, in particular, by source descriptions which might contain real user names, e-mail addresses, telephone numbers, geographic information, application names, private extensions, etc.

Vat provides several different audio encoding schemes such as PCM (pulse code modulation), ADPCM (adaptive differential PCM), GSM (general special mobile), and LPC (linear predictive coding). Privacy is supported by encryption. Each participant has to know the session key in advance and type it into the corresponding box of the vat menu.

The design of vat is based on the Light-Weight Session (LWS) principle described previously. In particular, it is an example of timing recovery by the receiver. Since the Internet may introduce delays varying over time, the vat receiver implementation tries to eliminate the jittering effect by delaying the playout time of the received audio samples. Delay jitter can be minimized by playout buffers. Packets are stored in the playout buffer at the destination and are delivered to the application at the correct time. The required size of the playout buffer depends on the delay variation of the received packets. If all packets have nearly the same delay, the playout buffer can be small. However, if the packet delays differ significantly, the playout buffer size must be augmented. In this case, early packets have to be stored in the playout buffer, and the end-to-end delay increases. Therefore, it would be desirable to have large playout buffers only in situations where the packet delays differ significantly. Vat tries to minimize the delay by adapting the playout time to the network's delay characteristics. Since adapting playout times for each packet would result in audible gaps, the playout time is adapted at the start of a talk spurt only by changing the length of the silence interval.

Vat also provides statistical features allowing the user to monitor the network statistics such as loss detection and delay analysis. Vat is primarily used in interactive Internet audio conferences but also for radio broadcasts over the Internet.

Integration of vic and vat by a Conference Bus

Vic and vat can make use of the so-called conference bus on which both applications can broadcast messages. All applications receive a copy of those messages for which they have registered. The implementation of the conference bus is based on a computer-local IP multicast. The conference bus mechanism allows different conferencing tools of a computer to collaborate with each other. Vic and vat use the conference bus as follows.

- Vat broadcasts focus messages indicating the RTP CNAME of the current speaker. Vic monitors these messages and switches the viewing window to the person identified by the globally unique CNAME, using voice-switched video windows.
- The LBL MBone tools have the ability to mute or ignore media sources. Muting can be controlled by the conference bus. A moderator can give the floor to a participant by multicasting a floor directive message containing the participant's CNAME. Each receiver is then able to mute all participants except the floor holder.
- Each real-time application includes a buffering delay (playback point) to adapt to packet delay variations. The buffering delays are often adjustable by the different tools. Different playback points would result in unsynchronized playout. By broadcasting the different playback points across the conference bus, the different tools can compute the maximum value of all playback points and can all adjust to the same value, thus improving the intermedia synchronization.

7.3.3   Adaptive Applications

A severe problem of real-time applications running over the Internet is the unpredictable network behavior. Packet loss rates of 20% or more are no exception. However, such packet loss rates are not acceptable for real-time applications, in particular, for audio conferencing scenarios where some participants do not use their native languages.

There are two solutions to overcome this problem. The first one is to avoid packet loss due to network congestion by reservation of the required resources. This approach can be achieved by establishing dedicated ISDN or ATM connections. The Resource ReserVation Protocol (RSVP) [10] has been designed to support resource reservation in the Internet and to achieve better services than those of pure best-effort (see Chapter 5). Examples of real-time applications using resource reservation are given in section 7.3.5.

The second approach assumes that resource reservation in the Internet is not possible (as it is today) or that resource reservation is too expensive for many users. In this case, one must accept the varying Quality-of-Service (QoS) the Internet provides. To avoid high packet loss, it is possible to adapt the transmission parameters of the applications to the current network situation. For example, if the network is not congested, one can transmit audio or video data streams with the full required bandwidth to achieve the best audio or video quality. If the network becomes congested, the transmission rate is decreased in order to reduce the congestion and the packet loss ratio.

RTP-based applications can make use of RTCP to adapt the transmission parameters from RTCP's QoS status reports. Two of those applications, IVS and FreePhone, were developed by the RODEO (high-speed networks, open networks) research group at INRIA (Institut National de Recherche en Informatique et en Automatique).

INRIA Video Conferencing System (IVS)

IVS implements a congestion control algorithm by adapting its output rate to the network conditions [46]. IVS is based on the ALF implementation concept and uses H.261 compression to perform well over low bandwidth links. A macro block is used as the application data unit (ADU) according to the ALF concept. (See Chapter 3 for details on H.261 video encoding.) Packets must begin and end on macro block boundaries and can consist of one or more macro blocks. The IVS packetization scheme allows a receiving application to decode ADUs independently from other ADUs. All information to decode a macro block is sent in a specific H.261 RTP header together with the H.261 data. The RTP payload type indicates H.261 encoding; the RTP timestamp encodes the sampling instant of the first macro block contained in the RTP data packet, based on a 90 kHz clock, which is a multiple of the natural H.261 frame rate of 29.97 kHz. The RTP marker bit is set to one in the last packet of a video frame.

A severe problem in lossy networks such as the Internet is the use of differential encoding schemes such as H.261 interframe encoding. To overcome this problem, vic does not use H.261 interframe encoding but uses intraframe encoding only, with the disadvantage of a lower compression ratio. IVS tries to recover from packet loss by applying two approaches. The first approach uses replenishment of macro blocks affected by packet loss. As soon as a receiver detects packet loss, it sends a negative acknowledgment to the sender, indicating which groups of blocks shall be replenished in the next image. The second approach periodically refreshes the image, using intracoding mode. The H.261 recommendation requires intra-encoding of each macro block at least once every 132 times it is transmitted. The IVS encoder can adapt the refreshment rate according to the packet loss information. IVS uses the negative acknowledgment approach in conferences with fewer than 10 users, and the intra- refreshment approach with more than 10 users.

A unique IVS feature is the end-to-end control mechanism adapting to the varying network situations. The end-to-end control mechanism is based on two components: a network sensor and a throughput controller.

The network sensor measures packet loss by RTP QoS reports generated by receiver after every 100 packets. These periodic QoS reports are more efficient than sending negative acknowledgments for lost packets as soon as the loss rate is higher than 1%, i.e., if more than 1 packet gets lost per 100 packets, which is nearly always the case in the Internet. The encoder performs control actions for every 100 encoded and sent packets. The throughput control algorithm adjusts the maximum output rate of the encoder so that the median loss rate over all receivers stays below a tolerable loss rate (e.g., 10%). In particular, the maximum output rate is decreased by a factor of two in that case. If the median loss rate falls below the tolerable value, the maximum rate is increased by a fixed fraction (e.g., 50%).

The output rate of the encoder can be adjusted by changing the frame rate, the quantization factors, or the motion vector search range. The details of these encoding parameters are explained in Chapter 3. IVS allows the selection of two different modes: privilege quality (PQ) and privilege frame rate (PFR). PQ is appropriate for applications requiring high precision. In this mode, the quantization threshold and the motion vector search range are constant and set to a value for maximal visual quality. The encoder waits a certain amount of time before encoding a new image. PFR mode is appropriate when movement perception is an important quality factor. The output rate is then controlled by using different quantization threshold and motion vector search range values.

The successor of IVS called Rendez-Vous [34] is currently being developed at INRIA. It already supports JPEG and H.261, and H.263 is to be implemented soon. MPEG-I and MPEG-II video files can be read by the tool and transcoded to one of the mentioned formats. Optimized scheduling procedures improve the system's resource usage, resulting in better runtime behavior and better quality rendered to the user. Rendez-Vous will support receiver-driven, layered multicast (cf. section 7.3.4) for audio and video flows. Figure 7–3 illustrates IVS in use.

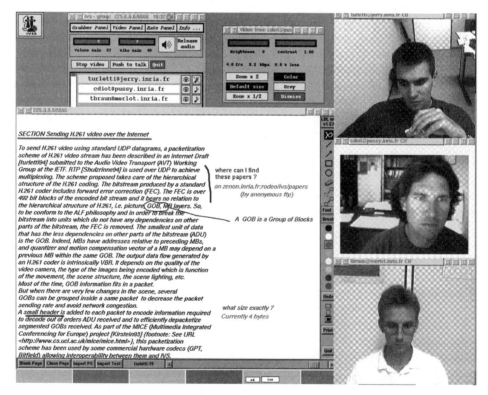

Figure 7–3 Multimedia Conferencing with IVS and wb

FreePhone

The audio component of Rendez-Vous is based on a new audio conferencing tool, called FreePhone, also developed by INRIA. FreePhone combines rate control and forward error correction to provide a stable audio quality over the Internet. Depending on the RTCP QoS status reports, the tool adapts the transmission rate and the amount of redundant information added to audio packets. Furthermore, FreePhone includes a mechanism to adapt the playout buffer depending on QoS statistics [8]. The overall architecture of the audio tool is shown in Figure 7–4.

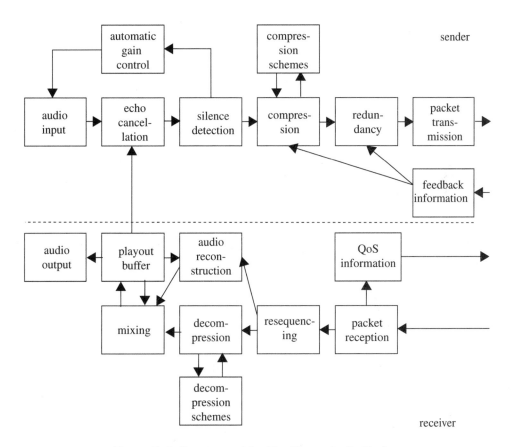

Figure 7–4 Structure of the FreePhone Audio Tool

The audio quality depends essentially on the number of lost packets and on the delay variations between successive packets. The problem of packet loss is solved by a combination of two approaches. Retransmission adds too much delay in wide-area networks with long round-trip times and is therefore inappropriate for real-time data transfer. FreePhone tries to recover from packet loss by adding redundant information to the audio packets. Simultaneously, the application takes care of avoiding network congestion, the main reason for packet loss.

Avoiding network congestion results in less packet loss due to congestion. However, reducing the transmission rate usually does not result in zero packet loss but decreases the packet loss only by a smaller degree. It may be better in certain conditions to add redundancy to the data stream than to decrease the transmission rate. Adding redundancy, however, results in a higher transmission rate, which in turn increases network congestion and may increase the loss rate.

The potential for FEC depends heavily on the characteristics of the packet loss process in the network. FEC mechanisms are more effective when the packets of the stream are dispersed throughout the data stream. The basic principle of FEC for multimedia streams was introduced in Chapter 5. FreePhone provides PCM encoding, ADM (adaptive delta modulation) encoding from 16 kbps (ADM2) to 48 kbps (ADM6), GSM (13 kbps), and LPC (4.8 kbps).

The combined rate and error control mechanism uses several levels of encoding combinations, described in Table 7–1.

	Combination	Bandwidth (kbps)
0	(PCM)	64
1	(ADM6, ADM3(1))	72
2	(ADM6, ADM3(2))	72
3	(ADM5, ADM3(1), ADM2(2))	80
4	(ADM5, ADM3(1), ADM2(3))	80
5	(ADM4, ADM3(1), ADM2(2))	72
6	(ADM4, ADM3(1), ADM2(3))	72
7	(ADM3, ADM3(1), ADM2(2), ADM2(3))	80
8	(ADM3, ADM2(1), ADM2(2), ADM2(3))	72
9	(ADM2, ADM2(1), ADM2(2), ADM2(3), ADM2(4))	80
10	(ADM2, ADM2(1), ADM2(2), ADM2(4))	64
11	(ADM2, ADM2(1), ADM2(3))	48
12	(ADM2, ADM2(1))	32
13	(ADM2, ADM2(2))	32

Table 7–1

Combination 3 means that ADM5 is used to encode the nth packet and that redundant packets n-1 and n-2 are included in the nth packet with encoding schemes ADM3 and ADM2, respectively. The control algorithm changes from combination i to combination i+1 if the global QoS is below a certain threshold value. If the QoS is above a certain threshold value, then the algorithm changes from combination i to i-1. For low levels of i, the algorithm does not decrease the bandwidth but increases the amount of redundant information. For large values of i, the algorithm is more a rate control than an error control mechanism. Experiments have indicated that the effective loss rate after packet reconstruction could be kept between 0% and 5% even if the network packet loss rate varied from 15% to 40%.

The global QoS measure represents the overall QoS of the audio received at all destinations. A so-called 90 percentile QoS is used to determine the global QoS. This means the global QoS value equals the smallest QoS better than 90% of the QoS values reported by the destinations. QoS reporting is again based on RTP receiver reports. This value might be problematic in conditions where the receivers are very heterogeneous. Issues of receiver heterogeneity are discussed in the next section.

7.3.4 Receiver Heterogeneity

The Internet is a heterogeneous network consisting of very different networks and end-systems. Private users often have low-bandwidth access, e.g., over telephone or ISDN lines, whereas large institutions, in particular, many research organizations and universities, have high-bandwidth access to the Internet via national research ATM networks. Also, the performance of the end-systems differs significantly. While PCs are familiar platforms for the user at home, high-end workstations with multimedia equipment are often found in the research or industrial environment.

However, it is often the case that users at very heterogeneous end-systems are connected to the same on-line session. For example, lectures are transmitted in the teleteaching project at Heidelberg-Mannheim via ATM between two lecture rooms at each university. The students in the seminar room or at a multimedia workstation of the university with hardware video decoders could receive the video, audio, and wb data stream with the full bandwidth of several Mbps and enjoy the best quality. Simultaneously, individual students can be connected via ISDN to the same session. The students at home are willing to pay at most for one or two 64 kbps ISDN channels to receive the video data at their PC with a software decoder.

Obviously, it is not a good approach to transmit the lecture with either the high bandwidth for the university users or with the low bandwidth for the students at home. Either the university users would have much worse quality than desired or the private user would have a very high packet loss resulting in unusable audio or video quality. Adaptive applications would calculate a supported rate somewhere between the high and the low bandwidth, which is a bad compromise in situations with very heterogeneous receivers. There are two approaches to overcome this problem.

- Application-level gateways can be inserted between the sender and the receivers, translating the original data stream to a bandwidth that the receiver is able to support.
- The data stream can be encoded hierarchically, using N different levels. Each receiver can decide, depending on his requirements, how many levels

he wants to receive. More levels lead to higher quality, but receiving the lowest level provides a reasonable result.

A/V Gateways

The video gateway (vgw) [2][3] is a translator for digital video streams that allows the transformation of a high-quality and high-bandwidth Motion-JPEG video stream to a low-bandwidth and low-quality H.261 video stream.

An advantage of the video gateway is that it allows the connection of non-multicast-capable end-systems to a multiparty conference. Also, two remote video gateways can be interconnected via unicast links. This feature avoids high time-to-live values that would otherwise be required in the IP packets.

The video gateway joins two RTP sessions and properly translates the data and control streams between them. Data forwarding is much simpler than control packet forwarding. Data must be translated according to rules defined by the media formats. Control packet forwarding is more difficult since RTCP packets contain QoS reports that are related to the specific medium. A translator cannot just forward RTCP packets but, in addition, must translate RTCP parameters such as packet counts, byte counts, timestamp information, etc. in a consistent way.

Vgw supports a limited set of video formats. JPEG, nv, or H.261 can be translated. The audio equivalent, agw (audio gateway), translates between PCM and LPC. Each input format is translated to an intermediate representation that is transformed to the output format. Combining arbitrary video input and output formats would require a true-color pixel representation as intermediate representation. This would lead to an unacceptable performance penalty. Therefore, only certain intermediate formats are supported, limiting the possible source and target combinations. For example, JPEG and H.261 use an intermediate format based on DCT coefficients for macro blocks. See Chapter 3 for details.

Figure 7–5 shows a scenario where two RTP gateways (vgw) interconnect two multicast-capable ATM networks over a point-to-point high-speed link. Each of the video gateways also connects low-bandwidth H.261 receivers to the multicasted data stream. The gateway approach has the advantage that standardized and available video codecs can be used.

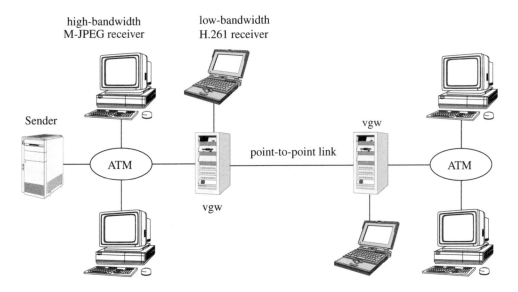

high-bandwidth
M-JPEG receiver

low-bandwidth
H.261 receiver

Sender

vgw

point-to-point link

ATM

ATM

vgw

Figure 7–5 A Video Gateway Scenario

Receiver-Driven Layered Multicast

Layered media streams are a more elegant approach supporting receiver hetero-geneity in multimedia transmissions to multicast groups. In this model, rather than distributing a single level of quality using a single network channel, the source provides multiple levels of quality simultaneously across multiple chan-nels. Each receiver individually adapts to its reception rate by adjusting the number of layers it receives.

The complete audio or video signal must be decomposed into a number of discrete layers. The receiver selects the number of desired layers based on its performance parameters (workstation performance, codec hardware) and its net-work access bandwidth. The more layers a receiver is able to receive, the better the overall audio or video quality.

With the RLM (receiver-driven layered multicast) scheme [36], each layer is transmitted to a separate IP multicast group. The receiver can then control the layers to receive by joining or leaving multicast groups. The IP multicast model has the advantage that data streams sent to a group are forwarded by the rout-ers only if there is a receiver farther down the tree interested in receiving the data stream. If there is no receiver interested in a particular layer, the RLM approach ensures that multicast routers do not forward the layers not needed downstream.

The RLM approach moves rate adaptation from the sender to the receiver. The source does not take an active role in the adaptation process and transmits each layer of its signal to a separate multicast group. The receiver runs a simple algorithm to adapt to the transmission. On congestion, layers are dropped or spare capacity layers are added. The receiver adds layers until congestion occurs and backs off to an operating point below this level. Occasionally, a receiver should test for available bandwidth to detect if bandwidth becomes available again and, if so, should add a layer.

To support RLM efficiently, the layered compression scheme should also provide low complexity and high error resilience. Unfortunately, the compression schemes in use today are not appropriate for layered encoding. New schemes have to be developed for this concept.

Figure 7–6 shows a layered stream with three receivers. Receiver 1 receives all three layers and combines them to a high-quality signal. Receiver 2 desires medium quality by receiving two layers, while receiver 3 is satisfied with even lower quality provided by a single stream. One can see that multicast router 1 has to forward only two layers since no downstream receiver is interested in all three layers.

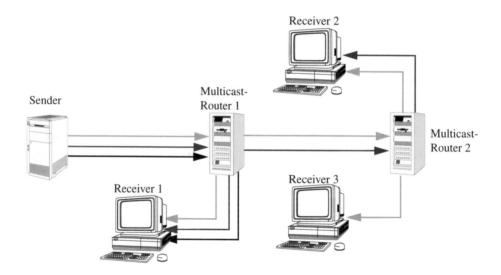

Figure 7–6 Receiver-Driven Layered Multicast (RLM)

7.3.5   Real-Time Applications with Resource Reservations

A/V applications running over the Internet can suffer strongly from network congestion. Even adaptive applications can only adapt as long as certain bandwidth and delay threshold values are achieved. Bandwidth reservation techniques can be used to guarantee a minimum level of QoS. However, the network must support bandwidth reservation, which is not the case in the Internet. Resource reservations may be supported in the Internet by the Resource Reservation Protocol (RSVP). But at this time, it is still not clear when RSVP will be globally deployed in the Internet.

The use of RSVP particularly makes sense when IP and RSVP are running over networks such as ATM or ISDN, which are able to use RSVP reservations for bandwidth reservation on their own level. Another possibility is to run A/V applications directly on top of the ATM or ISDN networks. There are already video conferencing products running directly over ISDN. Native ATM A/V applications are also being developed or are already available.

Mbone Applications over RSVP

Many MBone tools are available as source code, allowing researchers to perform experiments and extensions. Recently, RSVP has also been integrated in some experimental versions of conferencing tools such as vic. For example, PATH messages can be used by the sender of a video stream to indicate the required bandwidth for a good quality video transmission. The receiver can derive the QoS parameters of the reservation (RESV) message from parameters contained in the PATH message [12].

The RESV message may be used to reserve system, router, or network resources. The implementation [12] allows the receiving user to specify by a slider (the bottom one of the three sliders) how much bandwidth should be reserved for the video stream. The expected transmission bandwidth is selected by the sending user by another slider (the top one), is transmitted via PATH messages to the receiver, and is displayed in the RSVP information window. It can be used by the receiver to adapt his resource reservation (Figure 7–7).

In an ATM environment, the RSVP QoS parameters are mapped to ATM parameters, and dedicated virtual circuits with the corresponding ATM parameters are established by the underlying IP over ATM implementation. Then, the ATM virtual circuits provide the desired guaranteed QoS from the ATM network to the application data flow. In case of ATM, the receiving user regulates the bandwidth of the ATM connection by modifying the reservation slider.

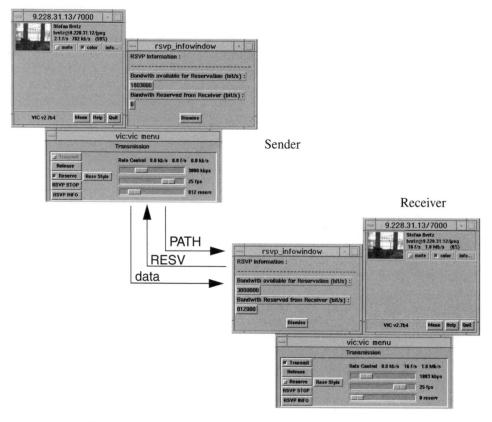

Figure 7–7 The vic Video Conferencing Tool over RSVP/ATM

Mbone Applications over Native ATM

The flexible, extensible, and object-oriented architecture of vic and vat supports heterogeneous environments and configurations. Vic and vat usually run over UDP and IP multicast, but their modular architecture allows them to be put on top of arbitrary communication systems such as native ATM.

Vic and vat have been extended to run over the native ATM socket interface of AIXv4.2 in addition to UDP/IP. The user can choose to connect to another user by UDP/IP or AAL5/ATM. If AAL5/ATM is selected, the ATM communication system establishes a best-effort (UBR) ATM connection, and the audio/video data is are exchanged over it [11]. To have QoS, the user can establish ATM VCs with QoS guarantees. To do that, the user opens a special ATM control menu from the standard vic user interface and selects the type of ATM connection to set up: unspecified bit rate (UBR), constant bit rate (CBR), variable bit rate (VBR), as well as additional traffic parameters such as the sustainable cell rate (SCR), peak cell rate (PCR), and maximum burst size (MBS). After these connection setup parameters have been specified, a new ATM connection is established and the old one is torn down.

7.4 Video Servers

In this section we look at the architecture of video server systems focusing communication related issues as and examine a video server for the MBone.

7.4.1 Architecture of Video Server Systems

A video server system consists of clients and servers connected to a high-speed network. In addition, *meta-servers* may also be part of a video server scenario [7].

Clients retrieve the video data from one or more servers. The video servers store the video data on high-volume storage devices such as disks or even disk arrays. Clients retrieve mainly video and associated audio data from the servers, but in addition, clients and servers also have to exchange control information to control the flow of the video data stream. Examples are the play, forward, or rewind functions. Figure 7–8 illustrates the video server architecture.

The meta-server provides control functions to manage the overall video server system. For example, a client can ask the meta-server for server names and addresses it needs to retrieve the video data stream. The meta-server may also provide information as file names, file size, frame rate, compression scheme, or video contents descriptions to the client. Depending on the information received from the meta-server, the client selects an appropriate video server from the set of possible ones.

The meta-server also controls the video servers themselves. It may configure the servers and manage the storage system, e.g., popular videos should be distributed to a larger number of servers than unpopular ones. The meta-server may also provide admission control functions and collect data required for billing, such as the number and duration of video retrievals. Access statistics collected by the meta-server may be used to optimize the performance of the overall system.

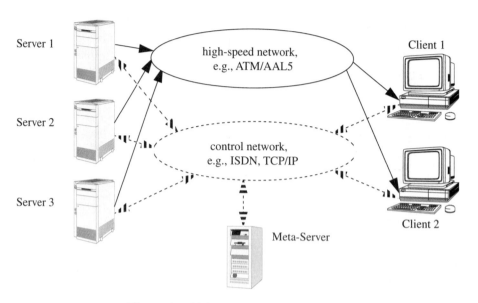

Figure 7–8 Video Server Architecture

Separation of Control and Data Transfer

The communication requirements for control and video data exchange may differ significantly. Video data will probably consume more bandwidth than control information, but the required reliability for control data may be much higher than for video data. Control data is exchanged bidirectionally, whereas video data flows in one direction from the server to the client only.

Therefore, it makes sense to use different communication systems to exchange control information and video data. For example, ATM networks could be used to transfer video data, while low-bandwidth links such as narrowband ISDN or telephone lines may provide sufficient capacity for control data. TV cable networks allowing unidirectional traffic to the clients are sufficient for video transfer, while bidirectional communication facilities such as telephone lines are required for control information. For video transfer, an unreliable AAL5/ATM service may be tolerable as long as packet loss is below a certain threshold, but 100% reliability as provided by a TCP/IP service is needed for control information exchange. The video server approach described in [27] uses this approach: AAL5/ATM for video transfer and TCP/IP for control information.

Server Arrays

An important issue in video server systems is QoS. Quality degradation may occur because of insufficient server or network performance. A single server can usually not support a large number of clients with a very large number of different video streams. Server arrays are required. The first approach distributes only complete movies to the servers. The client must ask the meta-server for the right video server to contact for retrieval of the desired movie. This approach scales as long as unloaded servers are able to satisfy certain requests. A meta-server should have the knowledge about the popularity of the movies and configure an appropriate number of video servers to store the most popular movies.

Another approach is to implement a striping mechanism [7]. In that case, all movies are divided into nonoverlapping subsets of video frames, and each frame subset is stored on a separate server. When a client requests a movie, each server transmits its frame subset as a substream. The client receives the different substreams and reconstructs them into the complete video stream. For example, it is possible to store all video frames with sequence numbers 3n-2 at video server 1, frames 3n-1 at server 2, and frames 3n at server 3 (n ε N). The approach has the advantage that the server load is well distributed among the servers and that each server must store the same amount of video data. However, the different servers must start their transmission approximately at the same time, and stream synchronization must be performed in order to avoid large buffering times at the clients.

Bandwidth Optimization

Another critical resource to be managed efficiently is network bandwidth. Video server applications are often based on bandwidth reservation in order to provide the required bandwidth to retrieve a movie with good quality. Resources can be reserved in advance, defining the desired time interval for playback. The meta-server then calculates the best starting time of a video transmission considering all other known requests and their bandwidth requirements.

In a video server scenario, the video server has some a priori knowledge about the stored video. This knowledge can be used to manage the required network resources efficiently. In particular, compressed video transmissions (e.g., based on MPEG compression) are very bursty. Peak bandwidth reservations can waste bandwidth since the peak bandwidth is required only for a very few scenes of a movie. An approach to avoid waste of bandwidth is to divide the video stream into different scenes with different peak bandwidth values. For each scene, a new reservation is performed. In [23], algorithms to calculate these time periods and the corresponding bandwidth parameters for ATM CBR and VBR services are described. The current implementation of this concept requires the tear-down and reestablishment of ATM connections. Future extensions of ATM signalling will probably support bandwidth renegotiation on the fly.

If a popular movie is requested by several clients simultaneously, these clients can join a single multicast group. The video server then may multicast the video stream over the high-speed network and use all advantages of multicast transmission. However, all clients of a multicast group receive the transmission simultaneously. It is not possible that a client interrupts the transmission for a few minutes and continues after the break. An approach to support such a scenario is to transmit the movie as N different streams with staggered starting times, e.g., stream n has a delay of 1 minute compared to stream n-1. This allows a client to interrupt the transmission for 1 minute. Before the break, the client received stream n-1; after the break it will receive stream n. This mechanism is also called near video on demand since it simulates a full interactive video-on-demand system. A near-video-on-demand system is able to support a very large number of clients by N video streams [33].

7.4.2   MBone VCR: A Video Server for the MBone

The MBone Video Conference Recorder (MBone VCR) [26] provides functions to record and play back MBone live sessions with multiple multicast multimedia data streams from different applications. A recorded session can consist of as many multicast channels as the user wants to record. The channels can originate from different locations and applications.

During recording, the MBone VCR synchronizes the multimedia data streams, based on information provided by RTP. To play back data streams, the MBone VCR transmits the recorded data, using the original timing and packet format. The receivers have only to run an application such as vic or vat to watch and listen to the recorded data.

The MBone VCR provides additional interesting features, such as indexing of the recorded data. The user can easily skip certain phases, for example, a certain speaker, of a session. Indices allow the user to jump back and forth between marked scenes. Other features, such as fast forward and rewind, are similar to those of regular VCRs. The user interface is similar to that of a regular video cassette recorder (Figure 7–9). Sessions to be recorded can be selected from the session directory (sd), which provides the user with lists of current and upcoming MBone events.

The MBone VCR allows users to watch and listen to conferences originating at locations with very different time zones. For example, an European user can record an MBone transmission from a scientific conference in Hawaii during the night and watch it the next day.

7.5 Applications Requiring Reliable Multicast

Reliable multicast data transfer is required in many different application scenarios such as shared editing and whiteboard applications, multicast chat, and multicast file transfer. Since TCP does not support multicast, all multicast applications must run on top of UDP, which provides an unreliable datagram service. According to the ALF concept, reliability functions for multicast are usually integrated into the applications. Examples are given in the following subsections.

7.5.1   Whiteboard

An important component of interactive multimedia conferences in addition to audio and video are distributed shared whiteboards for text and graphics. Wb was developed at LBL and is a widely used whiteboard tool that allows any conference participant to create pages and to draw on any page. Wb is often used in

Figure 7-9 MBone VCR: A Video Recorder for the MBone

multimedia conferences together with audio/video conferencing tools such as vic and vat but can also be combined with ivs as depicted in Figure 7-3 on page 202.

The wb tool requires relatively low bandwidth. A typical value is 64 kbps, which is in the same order of magnitude as for audio conferencing. Similarly to vic, vat, and ivs, privacy can be achieved by encryption of the multicast data. All participants allowed to join a session must know the key to decrypt the DES-encrypted data.

The whiteboard surface is separated into several pages, where each page typically corresponds to a new viewgraph in a lecture. Any member can create a new page and can draw on any page. A special mode called lecture mode prevents users other than the lecturer from creating pages and drawing. Each member is identified by a globally unique identifier.

After a page is created, any drawing operation wb is offering can be performed. Text and PostScript® files can be imported by wb. Wb is intended for discussion and annotation of documents; it is not a full drawing tool. A basic set of drawing operations such as lines, pointers, boxes, circles, eraser functionality, free-hand drawing, and text annotations is provided. A set of different colors can be used by the different participants to indicate which user did a certain drawing operation or annotation.

In contrast to audio/video applications, wb requires very high reliability, close to 100%. Retransmission requests and retransmission packets are generated by algorithms known as scalable, reliable multicast (SRM) [20] integrated into the application. The details of SRM were described in Chapter 6. SRM is a typical, ALF-based protocol framework. Only packet formats for repair request and repair messages are defined and used in the wb application. The application can decide which SRM repair requests should be processed with high or low priority. The whiteboard application gives the highest priority to repairs for the current page, middle priority for new data, and lower priority for previous pages.

A successor version of wb, called MediaBoard, is currently being developed at University of California at Berkeley. In addition to conventional drawing operations such as lines, rectangles, and ellipses, it also supports text, images (GIF and PostScript), animations, and embedded video streams. Like wb, MediaBoard relies on the SRM framework for reliable multicast [47]. MediaBoard is being developed within the MASH project, which aims to provide a flexible and extensible toolkit for implementing multimedia conferencing tools with an emphasis on easy and flexible code reuse [37].

7.5.2   Network Text Editor for Shared Text Editing

Network Text Editor (NTE) [25] is a shared editor designed for the MBone. Many people can edit the same document and even single text objects simultaneously. However, the users must synchronize their work to avoid conflicts when two users edit the same text block simultaneously. This synchronization could be done by audio conversation during a conference. Text blocks can also be protected by blocking against undesired editing.

NTE has been designed to achieve good performance and scalability for many users. Therefore, temporary inconsistencies, i.e., different views of the shared text by different users, are tolerated. Inconsistencies may result from packet loss, late joining of group members, and simultaneous changes by different users to the same text object, since NTE is based on the unreliable IP multicast service.

Several mechanisms have been implemented to discover inconsistent views. Periodically, each group member multicasts session messages indicating conference membership and containing a timestamp of the most recent modification seen and a checksum over all data. The rate of the periodic session messages depends on the number of session members in order to avoid too much control traffic. If the timestamp of another site is later than the indicated timestamp, the receiver can request all changes from the missing interval. The checksums are used to detect which blocks differ between two or more sites. In addition, the site that has most recently been active multicasts the timestamps of all objects changed recently. A receiver then requests the newer versions from the other site if it does not have a newer version to multicast to others.

7.5.3   MultiTalk

The UNIX talk application allows two Internet users to talk with each other via a text-based user interface. The terminal screen is divided into two parts, one part for each participant. If the first user types some text, it appears in the upper part of the screen, the text from the second user appears in the lower part. This scenario can be easily extended to more than two users to form a chat application. In this case, there is only one window, and the text of all users is displayed in the order in which it arrives at a server. Traditionally, these chat applications are based on TCP. A central server establishes a point-to-point TCP connection to each client and exchanges the data. This method means that the server must copy the data n times to distribute it to n participants via n TCP connections.

The MultiTalk [4] application permits text-based conversations among several users, based on IP multicast. Similar to the Mbone A/V conferencing tools, two ports are set up, one for sending and receiving text data, the other for exchanging control data. Since IP multicast is based on the unreliable UDP transport service, mechanisms to guarantee reliable text data exchange should be implemented within the MultiTalk application. MultiTalk currently does not provide a reliable service; its reliability depends on the reliability of the network.

7.5.4 Multicast File Transfer

Multicast file transfer can significantly save bandwidth in situations where exactly the same file must be transferred to more than one receiver. Example applications are as follows.

- Distribution of new software releases
- Distribution of system configuration data (e.g., router or switch configuration data)
- News broadcast
- WWW page replication and caching
- USENET news

A good example of multicast file transfer is the distribution of new software releases. If a server supports 200 clients by unicast ftp, 200 consecutive file transfers would be required. With multicast, only one file transfer is necessary, which reduces server processing load and network bandwidth. Multicast file transfer functionality is currently being integrated into existing file transfer protocols such as TFTP [19], or special multicast file transfer protocols, such as the ones described below, are developed.

The imm (image multicaster client) tool is a noninteractive, view-only software package allowing clients to receive images from a server and display them in the background of the desktop's screen. In most cases, weather or cartoon images are multicasted by the imm server. File transfer, however, has considerably higher reliability requirements than audio/video transmissions. The imm tool achieves reliability from a retransmission scheme adapted for multicast. Each file is transmitted in a number of cycles. First, all packets of the file are transmitted; then, in each missing packet recovery cycle, the packets negatively acknowledged by one or more clients are retransmitted. The missing packet recovery cycle repeats until all receivers positively acknowledge the reception of the entire file [14].

The Multicast File Transfer Protocol (MFTP) [38] consists of a control part (Multicast Control Protocol, MCP) and a data transfer part (Multicast Data Protocol, MDP). MCP announces multicast data transfer to be performed by MDP. The announcements consist of address information (private addresses) and a description about the type of files to be transferred. The announcements are multicasted to a public multicast address. The public address is either a well-known address or it can be queried by the clients from a dedicated server. The data is sent to the private address via MDP, which uses an algorithm similar to imm. Again, data transfer is divided into several phases. After the complete file is multicasted, only negatively acknowledged data are retransmitted in the subsequent phases.

Another popular application of multicast file or reliable multicast data transfer might be WWW page replication and caching. Very popular pages could be distributed with a multicast file transfer protocol. Replication of WWW pages could be initiated depending on the access frequency of the original page. In case of frequent accesses, the server of the original page could distribute the pages to proxy servers located between the original server and the clients. In that case, the replicated page would be transferred from one of the proxy servers instead of from the original server to the requesting client. This technique reduces delay if the proxy server is closer to the client, saves network bandwidth, and distributes the server processing load.

USENET news is currently distributed by means of uucp copy batches or NNTP (Network News Transfer Protocol) to transfer net news files from one news server to another one. Copying news files from one server to n others requires n copy operations. A multicast-based copy procedure could reduce the number of copy operations to one. An approach in that direction has been implemented in the Muse (Multicast USENET News) protocol. News articles are multicasted by a news server to others. However, the current protocol lacks a reliable multicast mechanism and can therefore only be used together with reliable backup news transfer mechanisms [32].

7.6 Multimedia Applications in the World Wide Web

The World Wide Web (WWW) is often called the multimedia part of the Internet. This statement is somewhat misleading. The basic browser technology based on HTTP (HyperText Transfer Protocol) does not allow retrieval of audio/video data in real time. Moreover, distributing and sharing web pages over multicast technologies is not supported. The following sections address these problems and discuss recent solutions.

7.6.1   Multicast Web Page Sharing

In a conventional web browsing scenario, a client connects to a server and retrieves web pages, including text and graphics, by establishing a sequence of unicast TCP connections. However, one could also think about using a web browser as a presentation platform. Presentations or interactive discussions could be based on a multicast distribution of web pages within a multicast group. In such a scenario, the presenter browses through a set of presentation web pages, and all other participants view the same web page as the presenter. Several research projects are currently addressing this problem.

Three different approaches are used to solve the problem. The first approach modifies a given browser and adds multicast functionality. In the second approach, Java™ applets are used to multicast data and to control the browser's display. The third approach is not limited to WWW browsers and allows sharing of X applications, based on multicast data distribution.

Multicast web page sharing can also be used to implement a pointcast mechanism that brings the information of interest to mass consumers instead of requiring users to fetch the information.

Multicast Extensions of WWW Browsers

Shared Mosaic [30] is an extension of the WWW browser, XMosaic, from NCSA (National Center for Supercomputer Applications). Several participants can surf together through the web. A shared Mosaic session is loosely controlled. Anyone can give a tour of preferred web sites and share web pages with others. Any user can click on a hyperlink or open web pages via hotlists, and the contents of the web pages are distributed to other participants by means of IP multicast. The loose control may lead to synchronization problems. Synchronization should therefore be supported by other means such as an audio tool allowing the participants to negotiate who controls the shared Mosaic session.

mMosaic [15] is also derived from NCSA's XMosaic. Any member of the multicast group can distribute or receive web pages. Each group member has a control window showing all session members. If a session member wants to see the current page of another member, he or she just has to click on the name of that session member.

Java Applets for Multicast Data Distribution and Browser Control

While WWW page distribution tools described above are tightly connected to a specific browser such as XMosaic, the approach implemented by mWeb [40] and WebCanal [31] is browser independent. Both tools are implemented as Java applets and are based on a reliable multicast protocol to distribute web pages to multicast groups.

mWeb acts as a kind of gateway between a WWW browser and the MBone. It provides functions for the distribution of WWW pages and precaching of files to be used in a session. mWeb saves all received pages on the local disk. When a presentation based on WWW pages is distributed over the MBone, the presenter sends dedicated display messages to the group members, indicating which of the cached pages the browser should display. If a page is found in the cache, it is requested by the local mWeb client. mWeb is based on an SRM-like reliable multicast protocol.

The WebCanal application runs as an HTTP proxy server for any WWW browser. WebCanal loads web pages when a browser has issued a request to load a web page. The document is then fetched by the WebCanal application as with a regular proxy server, but the document is also sent to the multicast group. A batch mode also allows the distribution of a set of predefined web pages. When a web page has been received, WebCanal requests the browser to display the most recently received web page or a list of received pages from which the user can select one page.

Application Sharing in the Interactive Remote Instruction System

The Interactive Remote Instruction (IRI) System [1] was designed at the Old Dominion University at Norfolk for distance learning. It integrates audio, video, shared applications, and several multiuser collaborative utilities within a single user interface.

IRI distributes the teacher's audio, video, and presentation to the students via IP multicast. Students can take notes of the presentation with special utilities, play back recorded courses at a later time, ask questions, and give presentations themselves. Students can also meet privately with the teacher or in small student groups.

An interesting feature of the IRI system is the application-sharing component, which makes use of IP multicast to provide good scalability characteristics. The X Tools sharing process (XTV) of the IRI system allows the sharing of arbitrary X applications between teacher and students, e.g., a WWW browser as a presentation tool. XTV intercepts the X application requests of the WWW browser and sends them to the local X server and the X servers of the students. The students have exactly the same browser view as the teacher. Only one person, either the teacher or one of the students, holds a token. Only the token holder can interact with the WWW browser, e.g., load web pages and follow links (see Figure 7–10).

The XTV processes exchange X packets, using IP multicast. However, the X applications such as WWW browsers require reliable underlying communication systems. Therefore, a reliable multicast protocol (RMP) [49] is required between the XTV process and the unreliable UDP/IP based multicast service [1]. This is a significant difference from other approaches implementing shared X applications which are running over the reliable but not multicast-capable TCP protocol. In contrast to the X application sharing component, audio and video are multicast over raw UDP/IP.

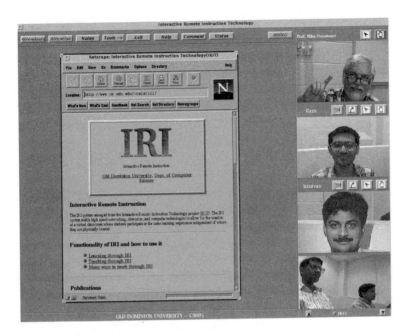

Figure 7–10 X Application Sharing

7.6.2 Audio/Video Streams in the WWW

The WWW technology and transfer of real-time data such as audio and video are among the most popular applications in the Internet. However, both technologies are based on different protocol architectures, making their integration difficult [9]. The HyperText Transfer Protocol (HTTP) [6] is based on TCP. Whenever a web page or a file is to be transferred, HTTP establishes a TCP connection between a client/server pair. Since TCP provides reliable and ordered delivery based on retransmissions, it is inappropriate for real-time data. Therefore, exchange of real-time data is typically based on RTP/UDP over the Internet.

A simple approach to transferring A/V data over the WWW is to download the complete A/V file from a web server via HTTP. After the A/V file has been downloaded by the client, a helper program is started to play the sound or the video locally. Obviously, this process introduces very large delays since A/V files can be very large and it often takes minutes to download them.

To reduce this delay, the output of the real-time data should be started as soon as the first data packets have arrived at the client. TCP should be avoided as the transport protocol for real-time data transfer. The basic approach to solving that problem is to exchange only control data over HTTP/TCP between the web client and the web server in order to retrieve initial information about the real-time data to be transferred. The web browser at the client machine then launches the A/V helper program (usually a browser plug-in) initialized with the parameters retrieved via HTTP from the web server. These parameters may include content information, the encoding type of the A/V data (e.g., MPEG video, PCM audio), or the address of the A/V server to contact for A/V retrieval.

The A/V helper program and the A/V server run a protocol to exchange control information required to start and stop the A/V transmission. For example, the Real Time Streaming Protocol (RTSP) [44] provides methods to realize commands—play, (fast) forward, (fast) rewind, pause, stop, and record—similar to the functionality provided by CD players or VCRs. RTSP can control either a single or several time-synchronized streams of continuous media. It can act as a network remote control for multimedia servers and can run over TCP or UDP.

The A/V data is received by a A/V client program, which may be identical to the helper program or may even run on a separate A/V client machine. The A/V client program is then able to output the A/V data as soon as it arrives at the client. The A/V data itself is transferred by a protocol other than RTSP or HTTP, for example, RTP/UDP. The protocol to be used can be negotiated with the A/V server via RTSP.

Using two different protocols for communication with the A/V server allows the redirection of the A/V server's output to another destination address different from the machine on which the A/V helper program is running. For example, via HTTP a client could retrieve from a web server A/V information such as the type and contents of an A/V file stored at a certain A/V server. The A/V helper program then instructs the A/V server to send the A/V data to a multicast address or a unicast address of another A/V client machine (see Figure 7–11).

For example, a teacher could ask a web server for interesting learning videos about a certain subject. The web server returns several addresses of appropriate video servers with the names or identifiers of the training material, such as A/V files, to the teacher. The teacher's helper program then contacts the given A/V server and instructs it to transmit the A/V file to the multicast group address, specifying the set of learners. RTSP can also be used to instruct an A/V server to record A/V sessions such as a lecture. In that case, the A/V server is ini-

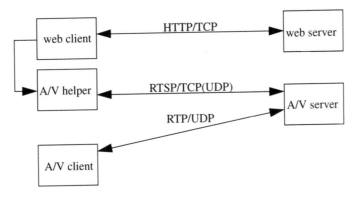

Figure 7–11 Real-Time Data Transfer over the WWW

tialized and instructed to record A/V data transmitted from an A/V client to a multicast address.

7.6.3 Conferencing Java Applets

Most of the currently available multimedia conferencing applications are available in binary or source code form for specific platforms, but many conferencing applications are not available on all platforms. The portability problem can be solved by Java implementations. A Java applet can be downloaded by a browser, and if a Java interpreter exists for the given platform, the user is able to run the program.

However, today only a few multimedia conferencing applications are written in Java. The Promondia tool [22] is a Java application, similar to wb, that implements chat sessions and a shared whiteboard. The JETS system of the University of Ottawa is a tele-collaboration system that enables sharing of a whiteboard, a text editor, and VRML applets. It also features an API so that a programmer can easily implement new collaborative applications in Java [45].

7.7 Interactive Multiplayer Games

Multiplayer games are relatively new IP multicast applications. Those interactive games have high requirements concerning both reliability and delay. The players must have complete and consistent views of the global game state, and the view must be updated quickly to allow the users to react very fast to changing games states.

The traditional architecture of a game implementation consists of a central server and n clients exchanging game data with the server. The server communicates with the clients, calculates the current state of the game, and distributes the relevant information to the clients, using reliable point-to-point connections.

A game implementation can make use of the IP multicast service by eliminating the server for distributing information to all clients. The clients can send information intended for all players to the multicast group (see Figure 7–12). Point-to-point communication between server and individual clients is mainly required to exchange control information such as registration, payment, etc. Since the reliability requirements for distributed games are high, TCP is often used for point-to-point communication. Reliable multicast mechanisms must be included in the applications to run on top of the UDP/IP multicast service (cf. Chapter 6).

7.7.1 On-line Casino

The implementation of an on-line casino [39] with n different black jack tables is based on the architecture depicted in Figure 7–12. Each of the n tables is implemented by a game server and several players. A separate multicast group address is required for each table.

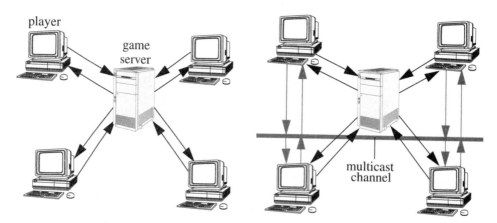

Figure 7–12 Distributed Game Implementation without and with
IP Multicast

The n tables are connected to a casino manager that can dynamically create and delete tables. The players interact only with the local game server of their table. The game server represents the card dealer of the table and the banker of the casinos who sells chips. The user selects one of the offered tables after entering the game. He joins the table, buys chips, and joins the game. After the game, he sells the remaining chips and leaves the table.

The current implementation does not provide reliable multicast mechanisms and should not be used in lossy networks. A reliable multicast mechanism should be integrated to run the on-line casino over the Internet.

7.7.2 MiMaze

MiMaze [18] is a multiuser game where several players meet in a virtual place (maze). MiMaze is being developed by the RODEO group at INRIA and is based on concepts of distributed real-time simulation and virtual reality. The user is represented by an icon moving through a labyrinth. The user can control the movements of the icon from the keyboard. The goal of the game is to walk through the labyrinth and kill the icons representing the other game participants. The movements of the icons should be effected with a short delay (not more than a few 100 ms) after the user's control action. In addition to walking through the virtual space and shooting, the user can also talk with other users by the FreePhone audio tool. FreePhone is also used to generate and distribute game sounds.

MiMaze is based on the client-server architecture and is depicted in Figure 7–13. The server manages participants joining and leaving the session and distributes the configuration of the virtual space. The clients (players) distribute their actions by multicasting, calculate the complete game state, and display the state on the players' screen.

The difficulty with realizing such an interactive game is to distribute state information among the participants in real-time but with a reliability of 100% for certain state information. For example, collisions between icons must be calculated with a very short delay. The solution currently under investigation is forward error correction. Very important events, such as movements possibly causing collisions or shots, may be multicasted several times (e.g., 25 times) per second. This rate increases the probability that the other clients receive the information.

Figure 7–13 MiMaze

7.8 Summary

This chapter provided an overview of multimedia applications, in particular, applications exchanging real-time data, such as audio and video, as well as applications supporting group scenarios with several interactively collaborating users.

Many of these multimedia applications are only possible because new advanced technologies in the area of audio/video compression, multicast communication, and high-speed networks offering QoS have been developed during the last years. These technologies are important building blocks enabling a new generation of multimedia applications. Although we have presented applications developed in research institutes and universities, many multimedia products are already commercially available. Examples include A/V conferencing tools running over ISDN or the Internet and video server components to implement video-on-demand services.

However, the process of integrating different technologies is still going on, and the integration of the World Wide Web with the Internet's capability to transfer multicast and real-time data is not yet finished. Also, TV and WWW technology might be integrated if more PCs were connected via cable TV networks offering bandwidths in the range of several Mbps. TVs and VCRs will get IP addresses—which will only be possible with the larger IPv6 address space. A/Vcompression/decompression hardware will be integrated into PC processors, or compression/decompression can be done more efficiently in software when processors become even faster or get special multimedia instruction sets.

An important issue of future multimedia applications will be the simplicity of user interfaces. Complex user interfaces prevent acceptance in a large market. In particular, it is necessary to allow the integration of several building blocks into applications with a simple and single user interface. The IRI system is a good example for an integrated multimedia application consisting of several different building blocks. The building blocks should be easily reusable to allow quick development of other multimedia applications tailored for other purposes. Reusable software components and mechanisms allowing the building blocks to interact with each other will become important. We expect multimedia applications to become the driving force for the future of digital communication networks.

References

[1]   H. Abdel-Wahab, K. Maly, and E. Stoica, "Multimedia Integration Into a Distance Learning Environment," 3rd International Conference on Multimedia Modelling, Toulouse, November 1996.

[2]   E. Amir, S. McCanne, and H. Zhang, "An Application Level Video Gateway," In *ACM Multimedia*, San Francisco, November 1995:511-522.

[3]   E. Amir and S. McCanne, RTPGW: An Application Level RTP Gateway, URL: http://http.cs.berkeley.edu/~elan/vgw/README.html

[4]   B. Anderson, MultiTalk, URL: http://pipkin.lut.ac.uk/~ben/multitalk

[5]   A. Banerjea, D. Ferrari, B. Mah, M. Moran, D. Verma, and H. Zhang, "The Tenet Real-time Protocol Suite: Design, Implementation, and Experience," *IEEE / ACM Transaction on Networking* February 1996: 1-10.

[6]   T. Berners-Lee, R. Fielding, and H. Frystyk, *Hypertext Transfer Protocol HTTP / 1.0*, RFC 1945, May 1996.

[7]   C. Bernhardt and E. Biersack, "A Scalable Video Server: Architecture, Design, and Implementation," In *Realtime Systems Conference*, Paris, January 1995:63-72.

[8]   J.-C. Bolot and A. Vega-Garcia, "Control Mechanisms for Packet Audio in the Internet," In *INFOCOM'96*, April 1996:232-239.

[9]   J. Bolot and P. Hoschka, "Sound and Video on the Web," presented at 5th WWW conference, Paris, May 1996.

[10]   R. Braden, L. Zhang, D. Estrin, S. Herzog, and S. Jamin, "Resource Reservation Protocol (RSVP) - Version 1 Functional Specification," Internet Draft, draft-ietf-rsvp-spec-14.ps, November 1996.

[11]   T. Braun and A. Reisenauer, "Implementation of an Audio/Video Conferencing Application over Native ATM," presented at European Workshop on Interactive Distributed Multimedia Systems and Telecommunication Services, IDMS'97, September 1997.

[12]   T. Braun and H. Stüttgen, "Implementation of an Internet Video Conferencing Application over ATM," presented at IEEE ATM'97 Workshop, May 1997.

[13]   D. Clark and D. Tenenhouse, "Architectural Considerations for a New Generation of Protocols," *ACM SIGCOMM '90*: 200-208.

[14]   W. Dang, Reliable File Transfer in the Multicast Domain, August 1993, ftp://ftp.merit.com/net-research/mbone/mirrors/imm/imm_overview.ps.Z

[15]   G. Dauphin, mMosaic Informations, URL: http://sig.enst.fr/~dauphin/mMosaic/ index.html

[16]   S. Deering, *Host Extensions for IP Multicasting*, RFC 1112, August 1989.

[17]   L. Delgrossi and L.Berger, *Internet Stream Protocol, Version 2 (ST2+)*, RFC 1819, November 1995.

[18]   C. Diot and L. Gautier, The Multicast Internet Maze, URL: http://www.inria.fr/rodeo/MiMaze

[19]   A. Emberson, *TFTP Multicast Option*, RFC 2090, February 1997.

[20]   S. Floyd, V. Jacobson, and S. McCanne, "A Reliable Multicast Framework for Light-weight Sessions and Application Level Framing," *ACM SIG-COMM'95*, August 1995:342-356.

[21]   R. Frederick, "Experiences with Real-time Software Video Compression," 6th International Workshop on Packet Video, September 1994:F1.1-F1.4.

[22]   U. Gall and F. Hauck, "Promondia: A Java-Based Framework for Real-Time Group Communication in the Web," presented at 6th International World Wide Web Conference, Santa Clara, CA, April 7-11, 1997.

[23]   S. Gumbrich, H. Emgrunt, and T. Braun, "Dynamic Bandwidth Allocation for Stored VBR Video in ATM End Systems," presented at 7th International IFIP Conference on High-Performance Networking, White Plains, USA, April 1997.

[24]   M. Handley, The sdr Session Directory,
       URL: http://mice.ed.ac.uk/mice/archive/sdr.html

[25]   M. Handley and J. Crowcroft, "Network Text Editor (NTE) A Scalable Shared Text Editor for the MBone," presented at ACM SIGCOMM, Cannes, September 1997.

[26]   W. Holfelder, "MBone VCR - Video Conference Recording on the MBone," *ACM Multimedia'95*, November 1995:237–238.

[27]   IBM International Technical Support Organization, "IBM Networked Video Solution Over ATM Implementation," IBM Redbook SG24-4958, January 1997

[28]   V. Jacobson and S. McCanne, "vat - LBL Audio Conferencing Tool," URL: http://www-nrg.ee.lbl.gov/vat/

[29]   V. Jacobson, "Multimedia Conferencing on the Internet," presented at ACM SIGCOMM'94 Tutorial, London, August 1994.

[30]   V. Kumar, *MBone: Interactive Multimedia on the Internet*, Indianapolis: New Riders Publishing, 1995.

[31]   T. Liao, "WebCanal: a Multicast Web Application," presented at 6th International World Wide Web Conference, Santa Clara, CA, April 7-11, 1997.

[32]   K. Lidl, J. Osborne, and J. Malcolm, "Drinking from the Firehose: Multicast USENET News," *Proceedings Usenix Winter* 1994, San Francisco, CA, January 1994.

[33]   T. D. C. Little and D. Venkatesh, "Prospects of Interactive Video-on-Demand," *IEEE Multimedia* 1(3):14–24, Fall 1994.

[34] F. Lyonnet, Rendez-Vous, the next generation videoconferencing tool, URL: http://www.inria.fr/rodeo/personnel/Frank.Lyonnet/IVStng/ivstng.html

[35] S. McCanne and V. Jacobson, "Vic: A Flexible Framework for Packet Video," *ACM Multimedia '95*, San Francisco, November 1995:511–522.

[36] S. McCanne, M. Vetterli, and V. Jacobson, "Low-Complexity Video Coding for Receiver-Driven Layered Multicast," to appear in *IEEE Journal of Selected Areas in Communications*.

[37] S. McCanne, E. Brewer, R. Katz, L. Rowe, E. Amir, Y. Chawathe, A. Coppersmith, K. Mayer-Patel, S. Raman, A. Schuett, D. Simpson, A. Swan, T.-L Tung, D. Wu, and B. Smith, "Toward a Common Infrastructure for Multimedia-Networking Middleware," presented at 7th International Workshop on Network and Operating Systems Support for Digital Audio and Video, NOSSDAV '97.

[38] K. Miller, K. Robertson, A. Tweedly, and M. White, "StarBurst Multicast File Transfer Protocol (MFTP) Specification," Internet Draft of work in progress, draft-miller-mftp-spec-02.txt, January 1997.

[39] R. Oppliger and J. Nottaris, "Online Casinos," KiVS'97, Braunschweig, Germany, February 19–21, 1997:2–16

[40] P. Parnes, M. Mattsson, K. Synnes, and D. Schefström, "The mWeb Presentation Framework," presented at 6th International World Wide Web Conference, Santa Clara, CA, April 7-11, 1997.

[41] H. Schulzrinne, *Voice Communication Across the Internet: A Network Voice Terminal*, Research Report, University of Massachusetts, July 1992.

[42] H. Schulzrinne, S. Casner, R. Frederick, and V. Jacobson, "RTP: A Transport Protocol for Real-Time Applications," RFC1889, November 1995.

[43] H. Schulzrinne, "RTP Profile for Audio and Video Conferences with Minimal Control," RFC1890, January 1996.

[44] H. Schulzrinne, A. Rao, and R. Lanphier, "Real Time Streaming Protocol (RTSP)," Internet Draft of work in progress, draft-ietf-mmusic-rtsp-02.ps, March 1997.

[45] S. Shirmohammadi and N. Georganas, "JETS: A Java-Enabled TeleCollaboration System," *Proc. IEEE International Conference on Multimedia Computing and Systems*, Ottawa, 1997:541–547.

[46] T. Turletti and C. Huitema, "Videoconferencing on the Internet," *IEEE/ACM Transactions on Networking* 4(3):340–351, June 1996.

[47]   University of California at Berkeley: MediaBoard, URL: http://www-mash.cs.berkeley.edu/mash/proejects/mboard/mb.html

[48]   D. Waitzman, C. Partridge, and S. Deering, *Distance Vector Multicast Routing Protocol*, RFC 1075, November 1988.

[49]   B. Whetten, T. Montgomery, and S. Kaplan, *A High Performance Totally Ordered Multicast Protocol, Theory and Practice in Distributed Systems*, Berlin: Springer-Verlag, LNCS 938, 1994.

Index

audio gateway, 206
audio/video conferencing, 196, 197
available bit rate, 102

B

backward learning, 128
bandwidth manager, 85, 89
bandwidth renegotiation, 213
bandwidth reservation, 208
basic rate interface, 95
B-channel, 84
BECN, 92
bidirectional symmetry, 25
B-ISDN, 96, 99
bit error rate (BER), 22
bit errors, 28
bottlenecks, 26
Broadcast and Unknown Server,
     118
broadcast mode, 20
buffer capacity, 26
burst tolerance, 102
bursty traffic, 25

C

cable TV networks, 108
Carrier Sense Multiple Access with
     Collision Detection, 83
CBDS, 90
CBT, 147
cell-based interfaces, 98
cell delay variation, 102
cell loss ratio, 102
cell transfer delay, 102
channel synchronization, 82
classical IP over ATM, 113
coding
    entropy, 38
    hybrid, 38
    source, 38
Committed Information Rate, 93

Common Intermediate Format, 7
compression, 9, 35, 226
computer-supported collaborative
     work, 20
conference bus, 199
congestion, 26
congestion window (cw), 171
connectionless protocols, 28
constant bit rate (CBR), 24
continuous bit rate, 102
continuous media, 3
core-based tree, 147
CSCW, 21
CSMA/CD, 83, 84, 140

D

data compression, 8
datagram, 126
DC coefficient, 44
D-channel, 84
DCT
    basic functions, 42
    coefficients, 42
decoding, 7
delay, 22, 23
delay variation, 24
Demand Priority, 86, 105, 106
Differential Pulse Code
     Modulation, 45
discrete cosine transform, 42
discrete media, 3
distance learning, 20
distance vector multicast routing,
     146
distance vector routing, 130
distributed multimedia, 19
Distributed Queue Dual Bus, 90
DPCM, 45
DQDB, 90, 106
DUAL, 134
DVMRP, 146

Virtual Cafe, 21
virtual circuit, 93, 126
virtual circuit identifiers, 98
virtual path identifier, 98, 114
visual audio tool, 197, 199
VPI, 98, 114

W

wb, 214, 223
Whiteboard, 214
wiring concentrator, 111
World Wide Web, 2, 218, 226
WWW, 217

X

X.25, 92, 94, 106, 108
XTV, 220

Z

zigzag sequence, 43